Touch this!

CONCEPTUAL
Physics for EVERYONE

by Paul G. Hewitt

Addison
Wesley

San Francisco   Boston   New York
Capetown   Hong Kong   London   Madrid   Mexico City
Montreal   Munich   Paris   Singapore   Sydney   Tokyo   Toronto

Cover Credit: The photo of Debbie Limogan, Genichiro Nakada, and
Natalie Limogan was taken by David Vasquez

Cover Design: Lillian Lee and M. Chew

ISBN 0-321-05162-9

Addison
Wesley

www.aw.com/physics

Table of Contents

Acknowledgments

My thanks go to many friends whose suggestions and feedback contributed to this book. In alphabetical order they include my friend since school days Howie Brand, my lifelong friend Ernie Brown, Hawaii elementary school teacher and specialist Kathy Chock, Chicago teacher friends Marshall and Margaret Ellenstein, former City College of San Francisco student Chuck Hatchett, pen pal Marilyn Hromatko, and three Hilo condo friends, Jack Ott, Mary Ann Ott, and Praful Shah.

I thank Marlene Hapai, science educator at the University of Hawaii at Hilo, for providing me with two classes of elementary school teachers who red penciled early manuscript. These teachers include Susan Cabral, Jocelyne Comstock, Claire Fernandez, Lorena Ishimoto, Mark Littorin, Shawn Paiva, Marilyn Quaccia, Shirlene Sato, Dawn Shirota, Agatha Solywoda, Emma Smith, Judy Stenger, Marilee Takaki, Connie Wood, and Janice Yasutake. Also at UHH, I thank scientist buddies Richard Crowe, George Curtis, Walter Steiger, and Suk Hwang.

I am grateful to Ken Ford and Joseph Slisko, whose helpful suggestions for recent editions of my textbooks have influenced this book. I thank my tennis buddy Anthony DeMotta for helpful artwork suggestions. I thank close friend Helen Yan for lettering the illustrations. I'm also grateful to my long-ago friend, Mary Jew, for her drawings of Newton and Einstein that open Chapters Three and Nine.

My thanks also to Ron Pullins of Focus Publishing, who encouraged the writing of this book years ago. Ron was my science editor when the sixth edition of *Conceptual Physics* was published by Little, Brown and Company. Thanks to his efforts, this is my first non-textbook, which has since been acquired by Addison Wesley Publishing Company. I am grateful to Ben Roberts and Linda Davis at Addison Wesley for their support. Ron, Ben, and Linda are among the very nicest people I've met in publishing.

Physics Dictionary

Fact: Contrary to public opinion, a fact in science is not immutable and absolute, but is generally a close agreement by competent observers of a series of observations of the same phenomena. The observations must be testable. The activity of science is the determination of the most probable; there are no absolutes. Facts that were held to be absolute in the past are seen altogether differently in the light of present-day knowledge. For example, where it was once a recognized fact in some cultures that the earth was flat, today it is a recognized fact that the earth is round. We distinguish between a fact and a "truth." In science we don't seek "the truth," some immutable and absolute knowledge. Truth in this sense is left to religious or political people.

Scientific Hypothesis: An educated guess that is presumed to be factual only until demonstrated by experiments. When a hypothesis has been tested over and over again and has not been contradicted, it may become known as a *law* or *principle*.

Law: A general hypothesis or statement about the relationship of natural quantities that has been tested over and over again and has not been contradicted. When less encompassing, laws are known as principles. Laws of nature describe how nature behaves, as found by experience and, hence, are *descriptive* and *not prescriptive*. Generally, laws are valid only within a limited area of experience. Newton's law of gravity, for example, is all that is needed to get a rocket to the moon, but is insufficient to describe what occurs in a black hole.

Theory: A synthesis of a large body of information. Physicists, for example, speak of the quark theory of the atomic nucleus; chemists speak of the theory of metallic bonding; geologists subscribe to the theory of plate tectonics; and astronomers speak of the theory of the big bang. The criterion of a theory is not whether it is true or untrue, but rather whether it is useful or not—whether or not it can predict new aspects of nature that were not previously known. A theory can be useful even though the ultimate causes of the phenomena it encompasses are unknown. For example, we accept the theory of gravitation as a useful synthesis of available knowledge that relates to the mutual attraction of bodies. The theory can be refined or, with new information, it can take on a new direction. It is important to acknowledge the common misunderstanding of what a scientific theory is as revealed by those who say, "But it is not a fact; it is *only* a theory." Here there is confusion between theory and hypothesis. Many people hold the mistaken notion that a theory, like a hypothesis, is tentative or speculative while a fact is absolute.

Concept: The intellectual framework that is part of a theory. We speak of the concept of time, the concept of energy, or the concept of a force field. Time is related to motion in space and is the substance of the Theory of Special Relativity. We find that energy exists in tiny grains, or quanta, which is a central concept in the Quantum Theory. An important concept in Newton's Theory of Universal Gravitation is the idea of a force field that surrounds a material body. A concept envelops the overriding idea that underlies various phenomena. Thus, when we think "conceptually" we encompass a generalized way of looking at things.

Prediction: In the everyday sense, prediction has to do with what has not yet occurred, like whether or not the Bulls will win tonight's game, or whether or not it will rain next weekend. In science, however, prediction is not so much about what *will* happen, but about *what is happening and is not yet noticed*—like what the properties of a hypothetical particle are and are not. A scientist predicts what can and cannot happen, rather than what will or will not happen. The predictions that a stock broker makes and the predictions that a scientist makes are similar in some respects, yet different in general. The stock broker predicts the likelihood of a stock reaching a certain value by a certain time—a prediction based on a wide variety of measurable factors. But the prediction is not of something that already exists. When a scientist predicts that a certain reaction will occur when A and B interact, for example, the prediction is of a reaction that normally has always occurred in nature whenever A and B interact. Knowledge of the reaction is new; not the reaction itself (unless A and B have never before interacted).

Preface

Why learn about science—especially physics, which everybody knows involves considerable effort? What do we say to young people who ask this question? If we say scientific knowledge helps to get a better job, our answer doesn't touch prospective artists, those looking to careers in business, or those who are more oriented toward people-centered rather than things-centered careers. Such an answer certainly doesn't touch a receptive chord in those people whose economic well-being seems far removed from scientific knowledge. Suppose we say that learning science is interesting in its own right, and worthy of the effort required. This is entirely reasonable to a science teacher, who has long ago become enraptured by the spectacle of nature. For the average student, this answer competes with many other interesting disciplines that make the same claim. Youngsters and others simply don't have the time and energy to learn everything. The process of education requires learning some fields of study while passing others by. Since learning science is unquestionably an uphill task, and questionably relevant to the everyday world, it's no wonder that most students minimize science courses in their education. Until recently, less than 15 percent of high school graduates in America have taken a course in the most basic science—physics.

The compelling reason for learning science begins with identifying science for what it is—the *study of nature's rules*. We are a part of the natural world, and to be ignorant of nature's rules is absolute folly. We all acknowledge that one cannot participate meaningfully in a game without knowing its rules. To be ignorant of the rules of any game, whether it be a game of sports, a computer game, or simply a party game, precludes any meaningful participation. Any participation soon becomes boring. And so it is with nature. While some youngsters are fascinated with the study of nature around them, most others are bored with anything not fast, furious, and dazzling. They are bored because, having no grip on what is going on around them, they simply don't care. Their natural sense of wonder hasn't been nourished by their home or school experiences—they have not been stirred to appreciate that everything is connected, that the connections are revealed by rules that are well within their ability to understand.

Learning about nature begins with learning its simplest rules: the rules of how things move, what things are made of, what makes them hot or cold, and how they're all connected. These involve physics—the central rules of science—the rules that chemistry and molecular biology are based upon. These are the rules that belong in the educational mainstream for science and nonscience-oriented students alike. A student's world and all of nature are viewed differently when the rules of physics are seen to underlie their functions. A student's impressions of the moon are forever changed once the student looks at it through a telescope and is directed to see the shadows cast by the mountainous ridges surrounding its craters. A student's viewing of video shots of orbiting astronauts is different when the student understands that the astronauts are freely falling around the earth. A study of electricity and magnetism means more when a student knows that they connect to become light. A student's view of all facets of nature is enriched when it is understood that everything is connected by a surprisingly small number of comprehensible rules. Physics offers a student the exhilaration that comes in knowing deep subjects well.

People with a conceptual understanding of physics are more alive to the world, just as a botanist taking a stroll through a wooded park is more alive than most of us to the flora and fauna teeming therein. The richness in life is not only seeing the world with wide open eyes, but knowing what to look for. Pointing out what to look for is the role of poets, artists, and teachers. Only the attuned ear can appreciate a Beethoven symphony; only the palate that has been cultivated can appreciate a fine Bordeaux; and only the mind that is familiar with the rules of physics can fully appreciate the physical world. This appreciation can be contagious. How fortunate is the child who is influenced by those who are knowledgeable (and how sad for children influenced by the ignorant). A knowledgeable teacher can be akin to a fine chef who supplies savory and healthful ingredients to what might otherwise be a bland soup.

Students will put considerable effort into a field of study that is seen to be personally relevant. Until recently, physics has been principally taught as applied mathematics, oriented only to students possessing mathematical skills. The mathematical emphasis has been a roadblock to those without a mathematical bent (and in many cases has obscured physics for those quite comfortable with math!). Physics taught conceptually, in recent years, is changing this. Conceptual Physics now presents a ramp for beginning students, rather than a vertical wall.

Conceptual Physics is making its way into the educational mainstream. Much of this book is excerpted from both my high school and college textbooks (published by Addison Wesley Longman Publishing Company). This book has much teacher-to-teacher commentary not found in the textbooks, some of which is from classroom experiences, some from personal experiences outside the classroom, all sprinkled with my personal educational philosophy. I'm aiming at parents and teachers who didn't learn much physics in their own student days, who would like to fill in gaps of understanding or mend past fences, who will see this book as an opportunity to surmount a previous roadblock.

Whereas knowledge of the rules of any game is necessary for full appreciation, in this book I try to present the clearest explanation of the rules of mechanics—basic physics. In follow-up books we'll treat other parts of physics—matter, heat, waves, electricity and magnetism, light, radio-activity, fission, fusion, and relativity. With this knowledge you can better appreciate the physics that is all around you, and better teach what you've learned to the young ones. Then as they grow, they'll recognize the difference between science and junk science, stand a better chance of keeping an open mind and making logical decisions in life. Whether your students are small kids or young adults, like you, they'll find that physics is everywhere—in everything you see, hear, smell, taste, or touch. Let's begin!

Dedicated to...

Lillian Lee
for her love

Huey D. Johnson
for his encouragement

Burl Grey
for his inspiration

Jacque Fresco
for his vision of a better world

and
Muhammed Ali
for being the greatest

Being in Equilibrium

I figured I was one of the lucky ones in Saugus High School in Massachusetts, for I knew exactly what I wanted to do with my life. I wanted to win the national amateur flyweight boxing championship and then get into a life of cartooning. This was reinforced by my school counselor who told me that because of my talent for art I wouldn't have to take the more academic courses in high school. That's right: "Wouldn't *have* to." When I wasn't sparring with my next-door neighbor Eddie MacCarthy, a professional lightweight boxer, I was drawing comic strips. When Eddie went off to the Korean War, he left me under the wing of local boxing hero, Kenny Isaacs. Kenny guided me along to the New England Amateur Athletic Union's silver medal when I was 16, but a brain concussion suffered in a warm-up bout for the next year's try at the gold dampened my dreams of ring glory.

I also put my comic strip plans on hold when I met cartoonist Ernie Brown and followed him into silk screen printing. Ernie turned out to be a life-long friend, and designs the cover lettering and title pages to all my textbooks. Eddie MacCarthy was killed in Korea just before I was drafted into the army. As luck would have it, the Korean conflict came to an end the day I finished basic training. I never went to Korea. Instead I remained at Camp Carson in Colorado for my two-year stint. Upon discharge I remained in Colorado to prospect for uranium, a popular pursuit at the time. My brother Dave and Ernie Brown joined me, but both left after a half year with no success. Then, with another prospecting companion, Millie Luna, I found uranium, enough to keep me excited for some years, but not enough for commercial gain. What I did gain was marriage to Millie three years later in Miami, Florida.

1

I went to Florida with Ernie Brown to seek silk-screen printing employ-ment. Instead I found a job painting billboards for an outdoor advertising company. Once employed, Millie joined me and we were married. It was in Miami that I met signpainter Burl Grey, who inspired me toward a new path—the quest for scientific knowledge. Burl was an extraordinary guy, with whom I had lost contact until very recently (as this book goes to press!). None of the painters wanted to paint with him because of his intellectual bent. Painting billboards was a two-person task, and conversations with Burl didn't involve the usual discourse about sports, cars, and sex—the fare of most painters. Rather than superficially discussing these topics, Burl preferred to ponder the hows and whys of the world about us and to query philosophical matters. He had no background in physics, probably missed an exposure to it in high school as I did, and he subsequently didn't know nature's rules. But he had an insatiable curiosity about the physical world.

I'll always remember one day when he called attention to the tension in the ropes that supported our weights and the weight of the staging we were on. He twanged the rope nearest his end and beckoned me to do the same with mine. He was interested in the relative tensions in the ropes. If we stood symmetrically and each had the same weight, the tensions should be the same in each rope. But Burl was heavier than I, and he wished to confirm that the tension in his rope ought to be greater than the tension in the rope nearest me. The twanging sounds confirmed his hypothesis. Together we reasoned it should be so, because more of the load was supported by his rope.

Such conversations drove many other painters bats, but I was enthralled. When I walked toward Burl, he asked if the tension in his rope increased as a result. We agreed it would, for we reasoned that his rope was supporting even more of the load. Burl then asked if the tension in my rope would be-come less. And we agreed that it would, for it would be supporting less of the total load.

We went further and used exaggeration to form our reasoning. If we both stood at an extreme end of the staging and leaned outward, we could imagine the opposite end of the staging rising like the end of a see saw, with the rope going limp. Then there would be no tension in the rope. From this we had an explanation for the case of me walking toward Burl. We reasoned

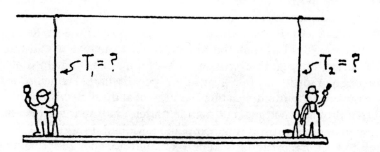

the tension in my rope would gradually decrease as I walked away from it. It was fun posing such questions and seeing if we could answer them. I'd bring the most interesting of these questions home after work and Millie and I would have a go at them. One that none of us could answer was whether or not the tension lost in one rope due to me walking away from it would be exactly compensated by the tension gained in the other rope. For example, if one rope underwent a 50 pound loss, would the other rope gain 50 pounds? Like exactly 50 pounds? And if so, would this be a grand coincidence? The answer to this wasn't known to me until more than a year later, when Burl's stimulation resulted in my leaving a full-time commitment to painting signs and going full tilt into formal education.

I returned to Massachusetts and made an appointment with the admissions director at MIT. I wanted to be a scientist and contribute to making a better world. I was convinced that the problems of the world would best be solved by scientifically-educated people, and I wanted to be one of them. When the director learned I hadn't taken science and math classes in high school, he advised me to go first to prep school and make up high school deficiencies in math and science. By this time Millie's and my first child was born. When the director further learned that neither I nor my family had any money to speak of, and that I had to work part time to support my wife and child, he suggested I go to the "state supported MIT," Lowell Tech, some 30 miles north. I took both his suggestions.

It was at Newman Preparatory School in Boston that I learned about the rule that provided the answer to the rope tensions problem! The rule was $\Sigma F = 0$, which said that for any object in equilibrium, the sum of the forces (ΣF) acting on the object would equal zero. So on the staging, the upward tensions in the ropes would be exactly equal in magnitude to the downward forces, our weights and the weight of the staging. The upward forces, however they may have varied, would always add up to our weights and the weight of the staging. So, yes, a 50-pound increase in one rope would be accompanied exactly by a 50-pound decrease in the other.

How different one's thinking is when one has or does not have a model to guide it. Burl was a technocrat of sorts, but if he had been mystical in his thinking, and guided me in that direction, we might have been more concerned with how each rope "knows" about the condition of the other. Such an approach, which intrigues many people with a nonscientific view of the world, could have led me to who knows what. I shudder to wonder if I could have become a pseudo-scientist of some sort had I been as greatly influenced by someone with flakier views than Burl's at this impressionable stage of my life. I owe a lot to Burl Grey.

So now I'm teaching others about what I didn't know when my thirst for scientific knowledge was whetted. It was exciting to discover that all the diverse phenomena of the world are tied together by a surprisingly small number of rules, and satisfying to share that excitement now. Everything is connected to everything else, and in a beautifully simple way. The rules of nature are what the study of physics is about. A knowledge of physics changes the way we see the world. There is sense to it all. And to point out this sense to my students is a very satisfying occupation—for me, much more stimulating than painting signs. So let's talk more about the equilibrium rule: $\Sigma F = 0$.

Equilibrium

Most things around us are in equilibrium. By that I mean they aren't undergoing any *changes* in motion. Things at rest, like the sign-painting staging just discussed, are in equilibrium—actually mechanical equilibrium, for there are other types of equilibrium, like thermal equilibrium, that we'll discuss much later. For now when we talk of equilibrium, we mean mechanical equilibrium. And soon enough we'll see that mechanical equilibrium is not confined to things at rest, for things that move at steady speeds without changes in motion are also in equilibrium. The rule of nature for objects at rest or moving equilibrium is that all the forces that act on the object balance out to equal zero—always. That's what we mean when we mathematically say, $\Sigma F = 0$.

A book lying motionless on a table is in equilibrium. Since the forces acting on the book balance out to equal zero, this means that the downward force of gravity on the book—its weight—must be exactly canceled by an equal upward force. The upward force is exerted by the table—the upward support force. This upward force is equal in magnitude to the weight of the book. The net force on the book is zero. If we call the upward force positive, then the downward weight is negative, and the two add together to become zero. Again,

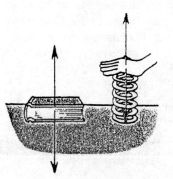

$\Sigma F = 0$. The net force that acts on any object in equilibrium equals zero—which is a powerful statement.

To better understand that the table pushes up on the book, compare the case of compressing a spring, as shown to the left. Push the spring down and you can feel the spring pushing up on your hand. Similarly, the book lying on the table compresses atoms in the table, which behave like microscopic springs, producing the support force.

When you step on a bathroom scale, the downward pull of gravity and the upward support force of the floor compress a spring that is calibrated to give your weight. In effect, the scale shows the support force. Since you are at rest, the net force on you is zero, which means the support force and your weight are equal in magnitude. For any object that is motionless, either no force is acting on it or there is a combination of forces that balance to zero. This idea that zero  net force acts on things in equilibrium is quite useful. For example, if you see that a force is exerted on something that doesn't move, then you know there is another force acting—one that is equal in magnitude and opposite in direction (or an equivalent combination of other forces). The net force on an object in equilibrium is always zero. This  simple rule was not known by Burl Grey. If he had known the rule, he could have answered the question about whether the gain in tension of one rope equaled the loss in tension in the other.

Question

Suppose you stand on two bathroom scales with your weight evenly divided between the two scales. What will each scale read? How about if you stand with more of your weight on one foot than the other?

When you hang from a rope, the atoms in the rope are not compressed, but are stretched apart. A tension force is produced in the rope. A rope under tension "twangs" if you pluck it, as Burl knew. If you hang at rest from a vertical rope, then the tension in the rope equals your weight. The rope pulls up and the force of gravity pulls down. Hang from two vertical ropes, and the sum of the tensions equals your weight. You're in equilibrium.

Answer

The reading on both scales adds up to your weight. This is because the sum of the scale readings, which equals the support force by the floor, must counteract your weight so the net force on you will be zero. If you stand equally on each scale, each will read half your weight. If you lean more on one scale than the other, more than half your weight will be read on that scale but less on the other, so they will still add up to your weight. For example, if one scale reads two-thirds your weight, the other scale will read one-third your weight. Get it?

Question

Harry the painter swings year after year from his bosun's chair. His weight is 150 pounds and the rope, unknown to him, has a breaking point of 100 pounds. Why doesn't the rope break when he is supported as shown to the left? One day, Harry is painting near a flagpole, and, for a change, he ties the free end of the rope to the flagpole instead of to his chair as shown to the right. Why did Harry end up taking his vacation early?

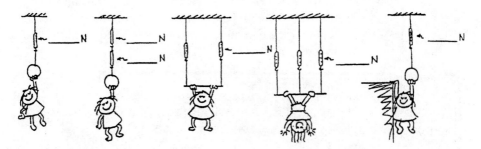

Consider the configurations above of Little Nellie Newton who aspires to be a gymnast. Since she hangs in equilibrium in each situation, the net force on her in each situation is zero. This means the upward pull of the rope(s) equals the downward pull of gravity. Suppose she weighs 60 pounds. You can figure what the scale readings are.*

Answer

In the left figure, Harry is supported by two strands of rope that share his weight (like standing on a pair of bathroom scales). So each strand supports only 75 pounds, below the breaking point. Total force up supplied by the strands of rope equals weight acting down, producing a net force of zero and no change in motion. In the right figure, Harry is now supported by one strand, which for Harry's well-being requires a tension of 150 pounds. Since this is above the breaking point of the rope, it breaks, changing his vacation plans.

---

* 1) 60 lb; 2) 60 lb, 60 lb; 3) 30 lb; 4) 20 lb; 5) 30 lb.

Vectors

When you were a child, perhaps you found that you could hang by a piece of vertical clothesline without breaking it. But you may have found that when the clothesline was strung horizontally, it would break if you attempted to hang from it. Similarly, you can't break a guitar string by pulling its ends with your bare hands, but when it is tightly strung and you pluck sideways on it, a small displacement to the side can cause it to snap. We see a great difference between applying forces along a string and at angles to the string. To understand this we consider an idea elucidated only a century ago by Oliver Heaviside in England—vector analysis.

A vector describes any quantity that has both magnitude and direction. Force is a vector quantity. By magnitude, we mean the strength of the force. Direction is as the name implies. Force, simply defined as a push or pull, can be represented by an arrow. When the arrow is drawn to scale, the length representing its magnitude and the arrowhead showing its direction, the arrow is called a *vector*. More specifically, it's a force vector, for we'll see later that other quantities such as velocity are vector quantities also. Vectors, like pictures, are often more descriptive than words. We'll touch lightly on vectors here and return to them again in the next chapter.

The vector shown below is scaled so that 1 centimeter represents 10 pounds; it is 3 centimeters long and points to the right. It therefore is a 30-pound force vector oriented to the right. If you pushed on an object with a 30-pound horizontal force, you could represent your push with this vector.

Suppose a friend also pushed on the same object with a 40-pound force in the same direction. Then the force on the object would be 70 pounds. Or if your friend pushed in the opposite direction, the resulting force would be 10 pounds in the opposite direction. So for parallel vectors, the rule for combining is simple: when in the same direction, the vectors add. When in opposite directions, they subtract. The resulting vector is called the *resultant*.

$$\underset{30}{\longrightarrow} + \underset{40}{\longrightarrow} = \underset{70}{\longrightarrow}$$

$$\underset{30}{\longrightarrow} + \underset{40}{\longleftarrow} = \underset{10}{\longleftarrow}$$

But what of vectors that act at angles to each other? The resultant is found by an intriguing technique called the parallelogram rule.* The rule states that we draw the pair of vectors so their tails coincide. Then we construct a parallelogram where the vector pair make up adjacent sides. The diagonal of

---

* A parallelogram is a 4-sided figure having opposite sides parallel, like the one shown here, or like a square or rectangle.

the constructed parallelogram represents the resultant. This is shown in the sketches that follow where the parallelogram is a rectangle.

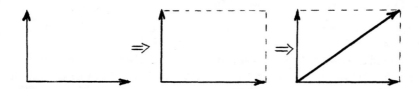

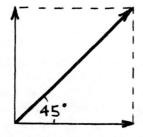

There is a special case of the parallelogram which often occurs. When two vectors that are equal in magnitude and at right angles to each other are to be added, the parallelogram becomes a square. Since for any square the length of a diagonal is $\sqrt{2}$, or 1.414, times one of the sides, the resultant is $\sqrt{2}$ times one of the vectors. For example, the resultant of two equal vectors of magnitude 10 acting at right angles to each other is 14.14.

On the facing page is a practice page that guides a student to seeing how the tension builds up in a pair of ropes that support a suspended weight. I've shown the constructions in the first two sketches of Question 1. Can you do the other two? And I've done the construction in the first sketch of Question 2. Can you do the other three?

In this practice page the resultant is the same for all sets of ropes in Question 1. The size of the resultant equals the weight of the ball. In Question 2 the process is done in reverse. Construction of parallelograms show the relative tensions in the supporting ropes. If you were careful, you found the rope tensions in the lower-right set-up as about three times the weight of the ball. As the ropes point closer to the horizontal, can you see how rope tension increases dramatically?

One of the most intriguing applications of the vector technique is the sailboat—especially showing how it is able to sail at an angle into the wind!

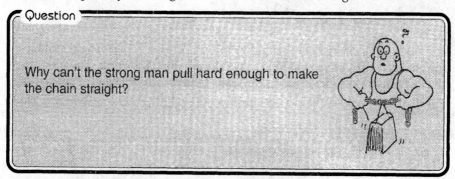

CONCEPTUAL **Physics** PRACTICE PAGE
Force Vectors and the Parallelogram Rule

1. The ball is supported in each case by two ropes. Tension is shown by the vectors. Construct parallelograms and find the resultant of each vector pair.

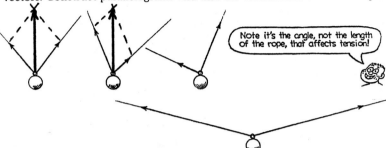

Note it's the angle, not the length of the rope, that affects tension!

2. Now do the opposite. Vectors **W** show each ball's weight. Dashed vectors show resultant of pairs of rope tensions (equal and opposite to **W**). Construct parallelograms and indicate rope tensions.

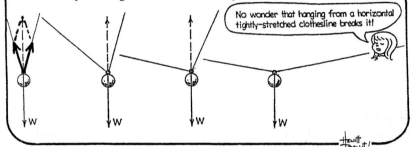

No wonder that hanging from a horizontal tightly-stretched clothesline breaks it!

Answer

For the book to hang in equilibrium, the tensions in each side of the chain must describe a parallelogram with a diagonal equal to the weight of the book—whatever the chain angles. As the chain is straightened, tension increases, and in the limit—if the chain were pulled straight—tension would be infinite. Clearly this can't happen. No matter how hard the ends are pulled, there must be a kink in the chain. Even a horizontally stretched guitar string sags slightly due to its weight, as can be detected with a thin laser beam.

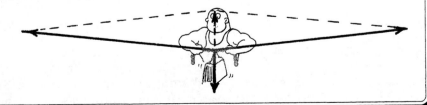

The Sailboat

Sailors have always known that a sailboat can sail in the same direction as the wind—downwind. The ships of Columbus were designed to sail principally downwind. Not until modern times did sailors learn that a sailboat can sail upwind (against the wind). It turns out that many types of sailboats can sail faster "cutting" upwind than when sailing directly with the wind. The old-timers didn't know this, probably because they didn't understand vectors and vector components. Luckily, we do, and today's sailboats are far more maneuverable than those of the past.

Let's see how. Consider the relatively simple case of sailing downwind. We show a force vector F due to the impact of the wind against the sail. This force tends to increase the speed of the boat. It is important to note that the faster the boat goes, the smaller will be the magnitude of F. That's because the speed of the wind relative to the sails decreases. For example, if the boat moves as fast as the wind, the relative wind speed is zero and no wind pushes against the sail. So a sailboat moving directly with the wind can sail no faster than the wind.

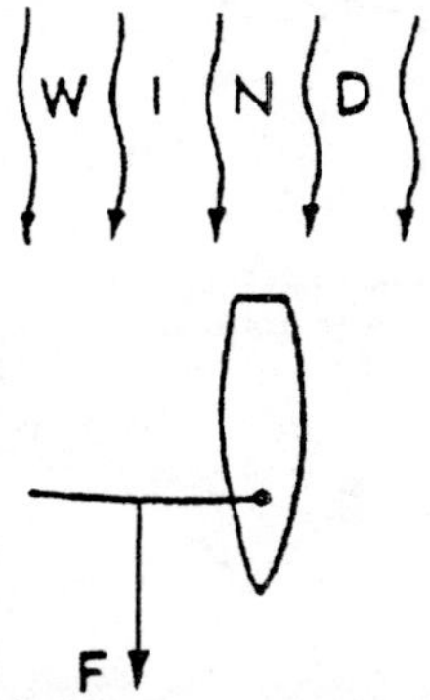

If the sail is oriented at an angle as shown below left, the boat will move forward but with less speed for two reasons. First, the force F on the sail is less because the sail does not intercept as much wind at this angle. Second, the force on the sail is not in the direction of the boat's motion. It is instead perpendicular to the sail's surface. Interestingly, whenever wind or any fluid (liquid or gas) interacts with a smooth surface, the force of interaction is perpendicular to the smooth surface. In this case, the boat will not move in the direction of F because of its deep fin-like keel (not shown), which knifes through the water and resists motion in sideways directions.

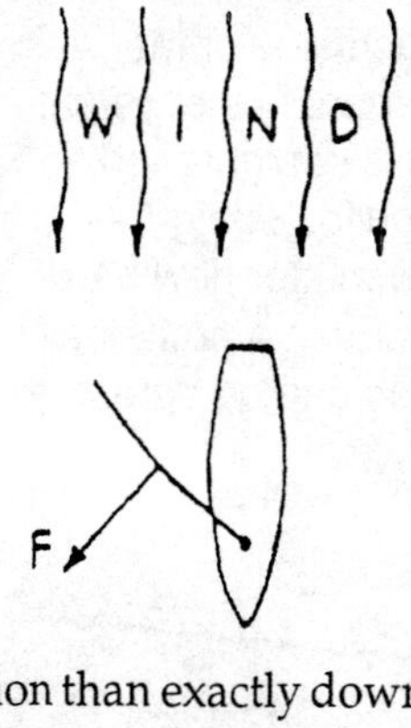

We can understand the motion of the boat by resolving F into perpendicular components, as shown in the figure on the left on the facing page. The important component is the one parallel to the keel and is labeled K. Component K propels the boat forward. The other component (T) is useless and tends to tip the boat over and move it sideways. The tendency to tip is offset by the heavy, deep keel. Again, maximum speed can only approach wind speed.

When a sailboat's keel points in a different direction than exactly downwind and its sails are properly oriented, it can go faster than wind speed. In the case of cutting across at an angular direction to the wind (center), the wind continues to move relative to the sail even after the boat achieves wind speed. A surfer, in a similar way, exceeds the speed of the

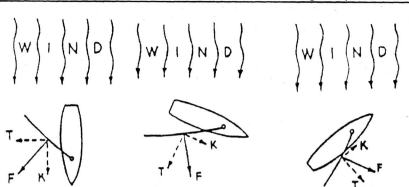

propelling wave by angling the surfboard across the wave. Greater angles to the propelling medium (wind for the boat, water wave for the surfboard) result in greater speeds. Can you see why a sailcraft can sail faster cutting across the wind than it can sailing downwind?

As strange as it may seem to people who do not understand vectors, maximum speed is attained by cutting into (against) the wind—that is, by angling the sailboat in a direction upwind (above right)! Although a sailboat cannot sail directly upwind, it can reach a destination upwind by angling back and forth in zigzag fashion. This is called tacking. As the speed increases, the relative speed of the wind, rather than decreasing, actually increases. (If you run outdoors in a slanting rain, the drops will hit you harder if you run into the rain rather than away from the rain!) The faster the boat moves as it tacks upwind, the greater the magnitude of F. Thus, component K will continue pushing the boat along in the forward direction. The boat reaches its optimum speed when opposing forces, mainly water drag, balance the force of wind impact.

Icecraft equipped with runners slide on ice and encounter no water drag. They can travel at several times wind speed when they tack upwind. Optimum speed is reached not so much because of resistive forces, but because the wind direction shifts relative to the moving craft. When this happens, the wind finally moves parallel to the sail rather than against it. We'll not go into detail about this complication or discuss the curvature of the sail, which also plays an important role.

In class I simulate a sailboat in the wind with a small cart, consisting of a block of wood with slots cut in it. Wheels allow the cart to move either back or forth, only in one direction as a keel constrains the motion of a sailboat. The sail consists of a square sheet of aluminum. When the sail is placed in the slot making it perpendicular to the direction of travel, no one is surprised to see the cart roll with the wind. Likewise when the sail is placed at an angle. But when I place the fan in front of the boat, directing the wind against the back side of the sail, the class is amazed to see the boat move forward, toward the fan. The sail doesn't "care" whether the wind impact is from the front or from the rear, as long as wind impact is against the back of the sail. This is one of my more impressive demos.

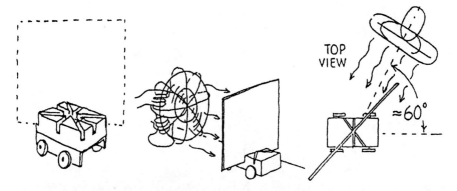

The central concept underlying sailcraft is the vector. It ushered in the era of clipper ships and revolutionized the sailing industry. Sailing, like most things, is more enjoyable if you understand what is happening.

Tension, Compression, Arches and Domes

We have seen that the tension in a stretched rope acts along the direction of the rope. Likewise, the tension between links of a taut chain lies along the direction of the chain. To a close approximation, this is true even if the chain sags. Consider the sag of a chain suspended between two points. The tension between links in the chain lies along the curve of the chain, which is why it takes the curved shape. The special shape of the curve is called a catenary.

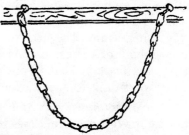

When we consider a stone arch, we see that compression rather than tension occurs. The stones of the arch press against one another, producing compression between adjacent stones. When a load is placed on a structure that is properly arched, compression strengthens rather than weakens the structure. This compression may or may not lie along the direction of the arch. The arches made by the ancients, as seen in elaborate aqueducts and other structures, were for the most part semi-circular in shape.

When the load being supported by an arch is sim-

ply its own weight, like the arch that graces the city of St. Louis, the shape of greatest strength is not the segment of a circle, but an inverted catenary. Then compression within it is everywhere parallel to the arch, just as tension between adjacent links of a hanging chain is everywhere parallel to the chain. You could make a secure arch out of slabs of slippery ice if its shape were a catenary. Modern arches, where strength is important, are usually catenaries.

If you twirled an arch through a complete circle, you'd have a dome. The weight of the dome, like that of an arch, produces compression. The catenary shape applied to domes was not appreciated by those who built domes and structures that required elaborate buttressing—the Notre Dame Cathedral in Paris, for example. One of the first successful domes is St. Paul's Cathedral in London, designed by Christopher Wren. Its shape? A catenary! Likewise for modern domes, such as the Astrodome in Houston.

How simple is the catenary concept. We can only wonder why it took so long for the architects and builders of stone structures in the past to discover this powerful idea. People have always wondered why it is so difficult to crush an egg by squeezing between one's fingers from each end of the long axis. Before you begin reading the next chapter, take a close look at the shape of an egg. One end is more rounded than the other. Hold the egg vertically and dangle a small chain beside it as in the picture on the facing page. Can you see that the chain follows the contour of the egg—shallow sag for the more rounded end, and deeper sag for the more pointed end? Nature has not overlooked the catenary!

Linear Motion

Physics courses traditionally begin with mechanics because it underlies all areas of physics. Although we've kicked off our treatment of mechanics with a chapter on equilibrium, mechanics traditionally begins with the study of *kinematics*—motion without regard to the forces that may accompany it. Kinematics is the study mainly of three concepts—speed, velocity, and acceleration. These are important concepts to learn, but not terribly exciting. In my teaching I define and distinguish between these three concepts and move on quickly to Newton's laws of motion, which are interesting, and which apply these concepts to interesting situations. Nonetheless, it is common for physics courses to spend considerable time and get bogged down in kinematics. Why? Principally for two reasons.

The first is that kinematics is rampant with catchy word problems—puzzles that delight teachers and very bright students. It's fun to feel like a genius and whip solutions onto the board. But these problems test wits more than knowledge of physics, and most are more about mathematics than physics. They intimidate less-than-bright students. Beginning a physics course this way, I feel, sends an unfair message about what physics is and contributes to low physics enrollments.

The second reason instructors spend so much time on kinematics has to do with a love affair with graphical analysis and its wonderful tools. Like some astronomers whose love is more for telescopes than stars, many teachers become more enamored with ticker timers, sonar ranging devices, and computers, than with the concepts they illustrate. I think students, given a choice between learning to plot motion graphs and learning about rainbows, would prefer rainbows. Nevertheless, kinematics usually gets overtime.

Talk to students who have recently taken a physics course and they'll likely be able to tell you the acceleration of free-fall. Now, ask them why the earth's interior is hot. I submit that hardly any have a clue. They don't know because that information was covered at the back of their physics book. Right now, before reading further, pause to answer the following: Why is lava red hot, and what heats the water in a thermal hot spring? I'm guessing that you, the reader, can't correctly answer these questions—and, I assume, for the same reason. When you took physics in school you probably spent so much time on the material covered in the front part of your textbook that you didn't get to radioactivity at the back. Or if you did, it was probably a hurried encounter. Radioactive decay in the earth's interior keeps the earth warm. Without it, the earth's interior would have cooled off long, long ago. In fact, before knowledge of radioactivity at the turn of the twentieth century, the famous physicist Lord Kelvin calculated the age of the earth to be about 25 million years. His figure was based on the temperature of the earth's interior as indicated by the temperatures of deep mines, of lava from volcanic eruptions, and the rate at which rock transmits heat. Kelvin's low figure was counter to the longer time speculated by geologists. We now know that radioactive decay in the earth's interior accounts for its high temperature after these 4.5 billion years.*

So let's get into kinematics and spend a brief chapter on it here. We won't get bogged down because we won't focus on computations, and there are no lab or computer activities with this book. Be relaxed about skimming some sections that don't seem to be your cup of tea, and digging into others.

Motion, after all, is present everywhere—from the vibrations of atoms in matter and the wigglings of enzymes in living cells, to the orbiting of planets around stars, which in turn swirl in galaxies that pivot the universe. Whether motion is simple or complicated, its underlying concepts are amazingly few in number. As mentioned, the most central are speed, velocity, and acceleration—what this chapter is about. We'll consider here the simplest kind of motion, linear motion—motion along a straight line. We'll make no attempt in this book to understand complicated things without first understanding the simplest things.

For our knowledge of motion we are indebted to the Italian scientist Galileo, born in the year of Shakespeare's birth and Michelangelo's death. Knowledge about motion at that time was dominated by the teachings of the Greek philosopher Aristotle, who lived nearly two thousand years earlier. Aristotle did not describe motion, but classified it: natural motion, which pro-

---

* Radioactivity is everywhere. Trace amounts are in the air we breathe, the food we eat, and the earth we stand on. Radioactive decay of traces of uranium and other minerals in common granite heat the rock by expending about 0.03 joules/kg per year. This isn't much energy in one year, which quickly dissipates from the rock at the earth's surface. But over a period of some 15 to 20 million years in the earth's interior, where dissipation is countered by the expended energy of surrounding rock, enough energy builds up to make the rock molten. Nuclear power, which has been around since time zero, heats the earth's interior.

ceeded from the "nature" of objects, and violent motion, which resulted from pushing or pulling forces. To Aristotle, it seemed natural that heavy things fell faster than light things; a rock, for example, naturally falls to the ground faster than a feather. Aristotle taught that the respective speeds of each happened at the instant of release—that a rock falls at one constant speed all the way to the ground, and a feather falls at a lesser constant speed. Aristotle assigned a special status to objects at rest. Objects normally at rest could only be made to move by the application of impressed forces producing violent motion. Up until Galileo's time there were great philosophical debates as to why objects moved as they did. Galileo did something very different from his contemporaries. Instead of concerning himself with why things move, his only concern, which he explicitly stated, was in how they move.

To measure motion Galileo used rates. A rate tells how fast something happens, or how much something changes in a certain amount of time. Rates that measure motion are speed, velocity, and acceleration.

Speed

Things in motion move certain distances in given times. An automobile, for example, may travel so many kilometers in an hour. **Speed** is a measure of how fast something moves—the rate at which distance is covered—always measured in terms of a unit of distance divided by a unit of time. In general,

$$\text{Speed} = \frac{\text{distance}}{\text{time}}$$

Any choice of distance unit and time unit can be used. For motor vehicles kilometers per hour (km/h) or miles per hour (mi/h or mph) are commonly used. For shorter distances, meters per second (m/s) are often useful. The slash symbol (/) is read as *per* and means *divided by*. Table 1 shows speeds in various units.*

Table 1 Approximate speeds in different units
25 km/h = 16 mi/h = 7 m/s
60 km/h = 37 mi/h = 17 m/s
75 km/h = 47 mi/h = 21 m/s
100 km/h = 62 mi/h = 28 m/s
120 km/h = 75 mi/h = 33 m/s

A car rarely moves at constant speed. On any trip the speed usually varies. So we speak of the **average speed**:

$$\text{Average Speed} = \frac{\text{total distance covered}}{\text{time interval}}$$

If we drive a distance of 50 kilometers, for example, in a time of 1 hour, we

* Conversion is based on 1 h = 3600 s, 1 mi = 1609.344 m.

find our average speed is 50 kilometers per hour. The same is true if we travel 200 kilometers in 4 hours:

$$\text{Average speed} = \frac{\text{total distance covered}}{\text{time interval}} = \frac{200 \text{ km}}{4 \text{ h}} = 50 \text{ km} / \text{h}$$

A distance in kilometers (km) divided by a time in hours (h) is an average speed in kilometers per hour (km/h). If we know average speed and time of travel, distance traveled is easy to find. A simple rearrangement of the above definition gives

Total distance covered = average speed × time

If your average speed is 50 kilometers per hour on a 4-hour trip, for example, you cover a total distance of 200 kilometers.

We may undergo a variety of speeds on most trips, so the average speed is often different than the speed at each instant—the **instantaneous speed**. This is the speed shown by the speedometer of a car. When we say the speed of a car at some instant is 60 kilometers per hour, we are specifying its instantaneous speed, and we mean that if the car continued moving at that speed for an hour, it would travel 60 kilometers. If it continued at that speed for half an hour, it would go half the distance, 30 kilometers. If it continued for one minute, it would go one kilometer.

Whether we talk about average speed or instantaneous speed, we are talking about the rates at which distance is traveled.

Questions

1. What is the average speed of a cheetah that sprints
   (a) 100 meters in 4 seconds?
   (b) 50 meters in 2 seconds?
2. A car traveled at an average speed of 60 km/h for 1 hour.
   (a) How far did it go?
   (b) At this rate, how far would it go in 4 hours? (c) in 10 hours?
3. During this 1-hr trip, is it possible for the car to attain an average speed of 60 km/h and never exceed a reading of 60 km/h on the speedometer?

Speed Is Relative

In a strict sense, everything moves—even things that appear to be at rest. They move relative to the sun and stars. A book that is at rest relative to the table it lies on is moving at about 30 kilometers per second relative to the sun. And it moves even faster relative to the center of our galaxy. When we discuss the speed of something, we describe its motion relative to something else. When we say an express train travels at 200 kilometers per hour, we mean relative to the track. When we say that a space shuttle moves at eight kilometers per second, we mean relative to the earth below. Unless stated otherwise, when we discuss the speeds of things in our environment, we mean relative to the surface of the earth.

Velocity

When we describe speed and the *direction* of motion, we are specifying velocity. Loosely speaking, we can use the words *speed* and *velocity* interchangeably. Strictly speaking, however, there is a distinction between the two. When we say that something travels at 60 kilometers per hour, we are specifying its speed. But if we say that something travels at 60 kilometers per hour to the north, we are specifying its velocity. A racecar driver is concerned primarily with speed—how fast the car moves along the track; an airplane pilot is concerned with velocity—how fast and in what direction the plane moves.

When something moves at constant velocity or constant speed, then *equal distances* are covered in equal intervals of time. Constant velocity and constant speed, however, can be very different. Constant velocity means constant speed with no change in direction. A car that rounds a curve at a constant speed does not have a constant velocity—its velocity continually changes as its direction continually changes.

Some physics instructors make a big deal about the distinction between speed and velocity. The distinction is important only when direction of motion makes a difference. When direction is important, velocity rather than speed is specified—positive or plus (+) when the direction is one way, say to the right or up, and minus (-) when the direction is opposite, say to the left or down. In solving velocity problems, distance is replaced by *displacement*—the straight-line distance from where you start to where you end up. If you make a round trip, the fact that you're where you started means your displacement is zero and hence, your average velocity is zero. So after a gruelling day at the Indy 500 where speeds approach 200 miles per hour, any car that returns to its starting place has an average velocity of zero. When this kind of physics is stressed in instruction, yuk! The distinction that *is* important, as we'll soon see, is that between velocity and *acceleration*.

Before progressing to acceleration, let's consider the "tricky" kinematics problem shown in the box below. This problem is in the class of questions I call "Next-Time Questions," for they can be used as a challenge question to end a class period. Then the next class can begin by discussing the answer, which enables a review of concepts before proceeding. Or they're posted in school hallways for a week or so, with answers displayed after students have had time to ponder them. We short-circuit this process here by displaying the answer on the next page.

CONCEPTUAL Physics

WHEN THE 10 km/h BIKES ARE 20 km APART, A BEE BEGINS FLYING FROM ONE WHEEL TO THE OTHER AT A STEADY SPEED OF 30 km/h. WHEN IT GETS TO THE WHEEL, IT ABRUPTLY TURNS AROUND AND FLIES BACK TO TOUCH THE FIRST WHEEL, THEN TURNS AROUND AND KEEPS REPEATING THE BACK-AND-FORTH TRIP UNTIL THE BIKES MEET, AND ⸘SQUISH! ⸘

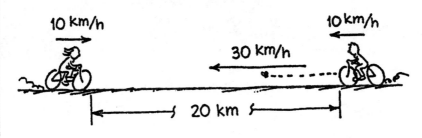

10 km/h   10 km/h

30 km/h

⊢——— 20 km ———⊣

QUESTION

HOW MANY KILOMETERS DID THE BEE TRAVEL IN ITS TOTAL BACK-AND-FORTH TRIPS ?

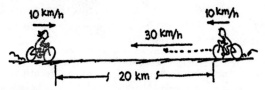

SOLUTION:

LET THE EQUATION FOR DISTANCE BE A GUIDE TO THINKING :

$$d = \bar{v}\, t$$

WE KNOW $\bar{v} = 30$ km/h, AND WE MUST FIND THE TIME t. WE CONSIDER THE SAME TIME FOR THE BIKES AND SEE IT TAKES 1 HOUR FOR THEM TO MEET, SINCE EACH TRAVELS 10 km AT A SPEED OF 10 km/h. SO,

$$d = \bar{v}\, t = 30 \text{ km/h} \times 1 \text{ h} = 30 \text{ km}$$

THE BEE TRAVELED A TOTAL OF 30 km.

The Usefulness of Equations in Problem Solving

The equations and formulas of physics normally provide the basis for solving problems. The key idea in solving a problem is to focus on exactly what is asked for, decide what concept of physics is at the heart of the problem, then write the formula that expresses this concept. The formula should then be re-expressed for the unknown term being sought. If the formula chosen is valid for the conditions of the problem, then each term of the formula is sought in turn. If they're given in the question, then the solution is simple—the problem is a simple "plug and chug" type exercise. Most often, the terms are only indirectly given, and one or more steps of repeating the initial process must occur.

Formulas guide thinking. They tell us what we must consider if confronted with a problem. Each unknown term presents a new step toward a solution. Often a term calls for another formula (or sometimes the same formula with different data). For example, in the bikes-and-bee problem in the box, you're looking for the distance the bee travels in its back and forth trips between the approaching bicycles. Write the formula for distance. It's $d = \bar{v}t$. You know the bee's speed $\bar{v}$, but you don't know the time it maintains this

THE KEY TO "TRICKY" SPEED AND DISTANCE PROBLEMS IS OFTEN TIME !

speed. There are a lot of things you don't know, such as how many back and forth trips the bee makes before it is squashed between the wheels of the bikes, and what the decrease in distance is for successive trips. But according to the formula, who cares? You only need time t to solve the problem, so you take a shortcut and concern yourself only with it.

Aha! A little thought tells you that the time for the bee is the same time it takes the bikes to come together, and that time you find easily with their speeds and the distance they travel. Voila! It's 1 hour. So t in your formula is 1 hour, which means the bee travels a total distance $d = \bar{v}t = 30$ km/h $\times 1$ hr $= 30$ km. If you didn't let the formula guide your thinking, you might have spent much more time and effort in attacking the problem.

It's common for many teachers to pooh-pooh the practice of reaching for a formula when a problem is presented. Plugging into formulas is seen as shortcutting a deeper understanding of physics. I disagree. For one thing, most students taking a physics course don't have the time or energy to develop the degree of understanding usually expected, which normally requires years of physics instruction. For another thing, I don't see that thoughtful use of formulas is counter to developing physics understanding at all. On the contrary, the formulas can be very useful guides to thinking. They show the connections between things and tell us what we can and can't ignore.

The important idea that *changing one thing changes another thing* is underscored with formulas. Change any term in a formula and another changes—always. Hence we can say: *you can never change only one thing.* This idea has wide application, currently with respect to the environment. Prevent water passage here and something else happens there. Remove trees here and something else happens there. The whole equation should be considered whenever we contemplate making changes.

Acceleration

When we teach the concept of acceleration, we make the academic plow setting deeper. Mainly because acceleration is confused with velocity. Velocity is a rate of change (distance/time), and acceleration is a rate of a rate (distance/time/time). Here's how we treat it in the classroom. When the velocity of a moving object changes, we have acceleration. We say the object accelerates. We accelerate when we change an object's speed, change its direction of motion, or by changing both speed *and* direction. By definition:

$$\text{Acceleration} = \frac{\text{change of velocity}}{\text{time interval}}$$

We link this formula to familiar examples, like acceleration in an automobile. When driving, we call acceleration "pickup" or "getaway"; we experience it when we tend to lurch toward the rear of the car. The key idea that defines acceleration is *change*. Suppose we are driving, and in 1 second we steadily increase our velocity from 30 kilometers per hour to 32 kilometers per hour, and then to 34 kilometers per hour in the next second, to 36 in the next second, and so on. We change our velocity by 2 kilometers per hour each second. This change of velocity is what we mean by acceleration.

$$\text{Acceleration} = \frac{\text{change of velocity}}{\text{time interval}} = \frac{2\ \text{km}\,/\,\text{h}}{1\ \text{s}} = 2\frac{\text{km}\,/\,\text{h}}{\text{s}} = 2\ \text{km}\,/\,\text{h}\,/\,\text{s}$$

In this case the acceleration is 2 kilometers per hour per second (abbreviated as 2 km/h/s). Note that a unit for time enters twice: once for the unit of velocity and again for the interval of time in which the velocity is changing. Also note that acceleration is not just the total change in velocity: it is the *time rate of change*, or *change per second*, of velocity.

Understanding speed and velocity is not really difficult; we all have some notion of fast or slow. Slow things have low speeds; fast things have higher speeds. Likewise for velocity. We can interchange the words speed and velocity in most cases without any trouble. But *acceleration* is something else—understanding the concept of acceleration is considerably more complex than understanding speed or velocity. Acceleration was first proposed and explained by Galileo in the 17th century. Until that time there was no clear way of describing changes in motion. And that's the key idea to acceleration—*changes in motion*.

The term acceleration applies to decreases as well as to increases in velocity. We say the brakes of a car, for example, produce large retarding accelerations—that is, there is a large decrease per second in the velocity of the car. We often call this deceleration, or negative acceleration. We experience deceleration when we apply brakes suddenly and lurch forward.

We accelerate whenever we move in a curved path, even if we are moving at constant speed, because our direction and, hence, our velocity is changing. We experience this acceleration as we tend to lurch toward the outer part of the curve. We distinguish speed and velocity for this reason and define acceleration as the rate at which velocity changes, thereby encompassing changes both in speed and in direction. The sketch shows three cases of acceleration.

Anyone who has stood in a crowded bus has experienced the difference between velocity and acceleration. Except for the effects of a bumpy road, you can stand with no extra effort inside a bus that moves at constant velocity, no matter how fast it is going. You can flip a coin and catch it exactly as if the bus were at rest. It is only when the bus accelerates—speeds up, slows down, or turns—that you experience difficulty.

When straight line motion is being considered, we can use speed and velocity interchangeably. When the direction is not changing, acceleration can be expressed as the rate at which speed changes.

$$\text{Acceleration (along a straight line)} = \frac{\text{change of speed}}{\text{time interval}}$$

Galileo developed the concept of acceleration by studying the motion of balls on inclined planes. His main interest was falling objects, and without suitable timing devices, the inclined planes effectively slowed the motion of the balls so he could investigate their motion more carefully.

What Galileo discovered was that a ball rolling down an incline picks up the *same amount of speed* in successive seconds: the ball rolls with an unchanging (constant) acceleration. For example, consider a ball that rolls down a particular slope and picks up a speed of 2 meters per second each second that it rolls. This gain per second is its acceleration. Its instantaneous velocity at 1-second intervals, at this acceleration, is then 0, 2, 4, 6, 8, 10, and so forth, meters per second. We can see that the instantaneous speed or velocity of the ball at any given time after being released from rest is equal to its acceleration multiplied by the time:

$$\text{Velocity acquired} = \text{acceleration} \times \text{time}$$

If we substitute 2 m/s/s for the acceleration of the ball, we can see that at the end of 1 second the ball moves 2 meters per second; at the end of 2 seconds, it moves at 4 meters per second; at the end of 10 seconds, it moves at 20 meters per second, and so on. After starting from rest, the instantaneous speed or velocity at any time is simply equal to the acceleration multiplied by the number of seconds it has been accelerating. Note that this relationship follows from the definition of acceleration. From $a = (\text{change in } v)/t$, simple rearrangement gives *change in* $v = at$ (when both sides of the equation are multiplied by t).

Questions

1. A particular car can go from rest to 90 km/h in 10 s. What is its acceleration?

2. In 2.5 s a car increases its speed from 60 km/h to 65 km/h while a bicycle goes from rest to 5 km/h in the same direction. Which undergoes the greater acceleration? What is the acceleration of each vehicle?

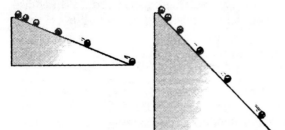

Galileo found greater accelerations for steeper inclines. A ball attains its maximum acceleration when the incline is tipped vertically. Then the acceleration is the same as that of a freely falling object. At lesser angles, the acceleration is a fraction of the acceleration of a freely falling object; when the incline is not tipped upward, and is horizontal, the ball rolls with no acceleration at all. Only friction would decelerate the ball. But getting back to free-fall, Galileo discovered that when air resistance is small enough to be neglected, regardless of weight or size, all material objects fall with the same constant acceleration. This was a remarkable discovery and is celebrated in the legend of his dropping objects of different masses from the Leaning Tower of Pisa.

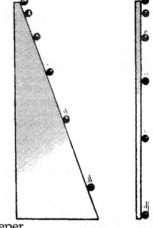

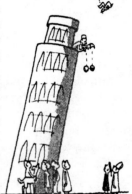

Answers

1. Its acceleration is 9 km/h/s. Strictly speaking, this would be its average acceleration, for there may have been some variation in its rate of picking up speed. Also, since acceleration, like velocity, is a vector quantity, its direction as well as magnitude should be specified.

2. The accelerations of both the car and the bicycle are the same: 2 km/h/s.

$$\text{Acceleration}_{car} = \frac{\text{change of velocity}}{\text{time interval}} = \frac{(65\ km/h\ -\ 60\ km/h)}{2.5\ s} = \frac{5\ km/h}{2.5\ s} = 2\ km/h/s.$$

$$\text{Acceleration}_{bike} = \frac{\text{change of velocity}}{\text{time interval}} = \frac{(5\ km/h\ -\ 0\ km/h)}{2.5\ s} = \frac{5\ km/h}{2.5\ s} = 2\ km/h/s$$

Although the velocities involved are quite different, the rates of change of velocity are the same. Hence, the accelerations are equal.

Galileo was a strong advocate of the notion that knowledge of the world is gained not by logic but by experimentation. Because he drummed this into the scientific world, Galileo is regarded as the father of not only modern physics, but of science in general. His discovery that a ball rolling along a smooth horizontal plane neither gained nor lost speed was to become Newton's first law of motion: an object once moving needs no force to keep moving, except to overcome any resistive forces present. This flew in the face of Aristotle's teachings, and opened the door to the sciences that followed.

Free-Fall—How Fast?

Things fall because of the force of gravity. When a falling object is free of all restraints—no friction, air or otherwise, and falls under the influence of gravity alone—the object is in **free-fall**. (We'll consider the effects of air resistance on falling in the next chapter). Table 2 shows the instantaneous speed of a freely falling object at 1 second intervals. The important thing to note in these numbers is the way the speed changes. *During each second of fall, the object gains a speed of 10 meters per second.* This gain per second is the acceleration. Free-fall acceleration is approximately equal to 10 meters per second each second or, in shorthand notation, 10 m/s² (read as 10 meters per second squared). Note that the unit of time, the second, enters twice—once for the unit of speed and again for the time interval during which the speed changes.

In the case of freely falling objects, it is customary to use the letter g to represent the acceleration (because the acceleration is due to *gravity*). Although the value of g varies slightly in different parts of the world, its average value is 9.8 meters per second each second or, in shorter notation, 9.8 m/s². We round this off to 10 m/s² in our present discussion and in Table 2 to establish the ideas involved more clearly: multiples of 10 are more obvious than multiples of 9.8. Where precision is important, the value of 9.8 m/s² should be used. Here we're using metric units. If we measure speed in units feet/second, then the acceleration due to gravity is 32 ft/s².

Note in Table 2 that the instantaneous speed or velocity of an object

Table 2 Free-Fall from Rest	
Time of fall (s)	Velocity Acquired (m/s)
0	0
1	10
2	20
3	30
4	40
5	50
.	.
.	.
.	.
t	$10t$

Question

What would be the speedometer reading on the falling rock 3.5 s after it drops from rest? How about 6 s after it is dropped? 100 s?

falling from rest is consistent with the equation that Galileo deduced with the use of his inclined planes:

$$\text{Velocity acquired} = \text{acceleration} \times \text{time}$$

The instantaneous velocity v of an object falling from rest after a time t can be expressed in shorthand notation as*

$$v = gt$$

The letter v symbolizes both speed and velocity. To see that this equation makes good sense, take a moment to check it with Table 2. Note that the instantaneous speed in meters per second is simply the acceleration $g = 10 \text{ m/s}^2$ multiplied by the time t in seconds.

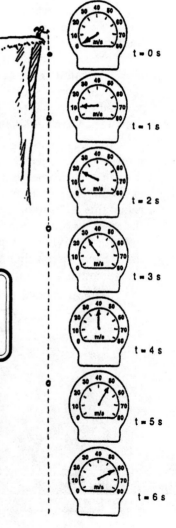

We get a good grasp on the concept of acceleration when we consider a falling object equipped with some sort of a speedometer. Suppose a rock was dropped from a high cliff and you could witness the drop with a telescope. If you focus the telescope on the speedometer, you'd note increasing speed as time progresses. This visual depiction of acceleration provides a way of looking at Table 2. Investigate it before reading further.

Physicists are wild about graphs. You can't find a physics book without them, including *Conceptual Physics*. A graph is a pictorial representation

Answer

The speedometer readings would be 35 m/s, 60 m/s, and 1000 m/s, respectively. You can reason this from Table 2 or use the equation $v = gt$, where g is replaced by 10 m/s^2.

---

* Note here an equation in pure letter form. Those with a math phobia who are somewhat intimidated by numbers are really challenged by equations of letters. But after all, they really are shortcut notations of the same relationships spelled out in words.

If instead of being dropped from rest, the object is thrown downward at speed v_o, the speed v after any elapsed time t is $v = v_o + gt$. We will not be concerned with this added complication here and will instead learn as much as we can from the most simple cases. That will be a lot!

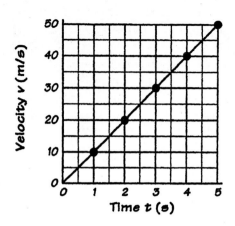

of an equation, or the data that fit the equation. Table 2 is shown here graphically. The straight line tells at a glance the direct proportion between velocity and time—double the time and the velocity is doubled, and so on. Graphs have their place and are very useful. But since they're a tool of physics rather than physics itself, we're not going to make a big deal about them.

So far we've considered objects moving straight downward in the direction of gravity. How about an object thrown straight upward? Here we'll not concern ourselves with the rock while being thrown, but only with the rock after it is released. Once released it continues to move upward for a while and then comes back down. At the highest point, when it is changing its direction of motion from upward to downward, its instantaneous speed is zero. Then it starts downward just as if it had been dropped from rest at that height. It is important to emphasize this point: it falls downward just as though it were simply dropped from the maximum height.

During the upward part of this motion, the object slows from its initial upward velocity to zero velocity. Its speed decreases—evidence of acceleration (or deceleration in the upward direction). How much does its speed decrease each second? It should come as no surprise that it decreases at the rate 10 meters per second each second—the same acceleration it experiences on the way down. So interestingly enough, as the sketch shows, the instantaneous speed at points of equal elevation in the path is the same whether the object is moving upward or downward. The acceleration is 10 meters per second squared the whole time, whether the object is moving upward or downward.

My teaching colleague Pablo Robinson makes this point with a handful of dimes. The game is that whenever an object loses a speed of 10 m/s, Mr. Robinson gives a student a dime (10 cents). And whenever the object gains 10 m/s, the

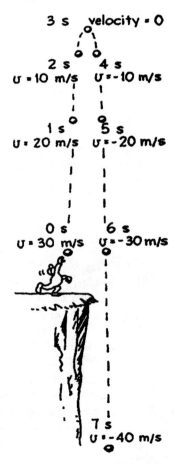

student gives Mr. Robinson a dime. They consider a projectile fired straight upward at 50 m/s. So what is the speed of the projectile 1 second after launch? The answer is 50-10 m/s, whereupon Mr. Robinson loses a dime. What is the speed after 2 seconds? Mr. Robinson loses another dime. It is soon seen that Mr. Robinson's net worth decreases 10 cents per second—for the first 5 seconds. He loses 5 dimes in the time it takes for the projectile to reach its maximum height. But thereafter, each second sees a gain of 10 m/s and Mr. Robinson collects a dime each second. There is a nice point made here: what was the change at the top of the path? The same 10 m/s! So the acceleration at the top of a trajectory is the same as elsewhere along the trajectory (assuming no air drag). This idea is better developed with examples of Newton's second law, as we shall soon see.

So if we toss a rock upward, and air resistance doesn't affect its motion, then the time up equals the time down to its starting elevation.

How Far?

How far an object falls is altogether different from how fast it falls. Using his inclined planes Galileo found that the distance a uniformly accelerating object travels is proportional to the square of the time. The distance travelled by a uniformly accelerating object starting from rest is

$$\text{Distance travelled} = \frac{1}{2}(\text{acceleration} \times \text{time} \times \text{time})$$

This relationship applies to the distance something falls. We can express it for the case of a freely falling object that starts from rest in shorthand notation as *

$$d = \frac{1}{2}gt^2$$

where d is the distance something falls when the time of fall in seconds is substituted for t and squared. If we use 10 m/s^2 for the value of g, the distance fallen for various times will be as shown in Table 3.

Note that an object falls a distance of only 5 meters during the first second of fall, although its speed at 1 second is 10 meters per second. This can be confusing if we think that the object should fall a distance of 10 meters. But to fall 10 meters in its first second of fall, it would have to fall at an average speed of 10 meters per second for the entire second. Note that it starts its fall at 0 meters per second, and its speed is 10 meters per second only in the last instant of the 1-second interval. So its average speed during this interval is the average of its initial and final speeds, 0 and 10 meters per second. To find the average value of these or any two numbers, we simply add the two numbers and divide by 2. This equals 5 meters per second, which over a time interval of 1 second gives a distance of 5 meters. As the object continues to fall in succeeding seconds, it will fall through ever-increasing distances as its speed continuously increases.

We return to the falling object equipped with the speedometer, which this time, has an odometer affixed to show the distance fallen. Note that the odometer readings are consistent with the data of Table 3. The last reading is

Table 3 Distance Fallen in Free Fall	
Time of fall (s)	Distance of fall (m)
0	0
1	5
2	20
3	45
4	80
5	125
.	.
.	.
.	.
t	$\frac{1}{2}10\,t^2$

---

* d = average velocity × time

$$d = \frac{\text{initial velocity} + \text{final velocity}}{2} \times \text{time}$$

$$d = \frac{0 + gt}{2} \times t = \frac{1}{2}gt^2$$

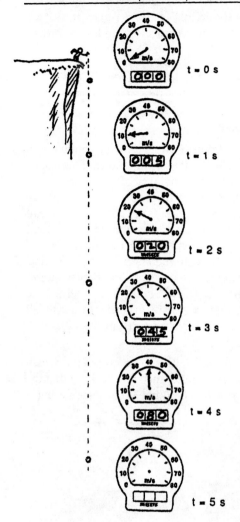

t = 0 s

t = 1 s

t = 2 s

t = 3 s

t = 4 s

t = 5 s

shown blank. Before reading further, fill it in yourself. Hooray for visualization when we learn concepts!

Table 3 on the previous page is shown graphically below. The curved line shows that between successive seconds the distance of fall increases—in each succeeding second, the object falls farther than the second before. Interestingly enough, if we instead plotted distance fallen versus time squared, a straight line would result. This is because distance is directly proportional to the square of the time. Again, we won't make a big deal of graphs.

Not all falling objects fall with equal accelerations. A sheet of paper falls altogether differently than the same sheet scrunched up. Or a feather certainly falls differently than a coin. But this is because of another factor—air resistance. In many science museums this is shown very nicely with a closed glass tube that contains a feather and a coin. In the presence of air, the feather and coin fall with quite different accelerations. But if the air in the tube is removed by a vacuum pump and the tube is quickly in-

verted, the feather and coin fall with the same acceleration. Although air resistance appreciably alters the motion of things like falling feathers, the motion of heavier objects like stones and baseballs at ordinary low speeds is not appreciably affected by the air. The relationships $v = gt$ and $d = 1/2\ gt^2$ can be used to a very good approximation for most objects falling from rest in air.

If you can make three correct predictions for the first of the following check questions and correctly answer

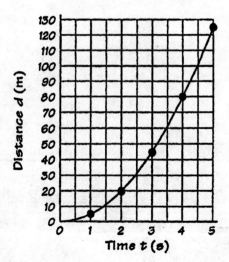

the second, you're in very good shape, and you should continue in confidence. If not, you have the choice of reviewing this material to get a better grasp, or conceding the fact you're not going to master all the physics presented here and continue onward to more fertile material.

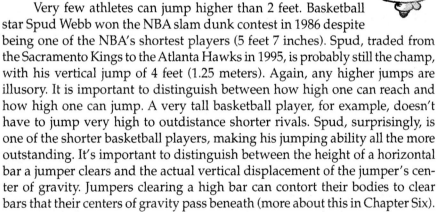

Questions

1. A cat steps off a ledge and drops to the ground in 1/2 second. What predictions can you make about the dropping cat?
2. If you drop an object, its acceleration toward the ground is 10 m/s². If you throw it downward, will its acceleration be more? Same? Less?

Hang Time

Some people, such as basketball players, track and field athletes, and ballet dancers are gifted with great jumping ability. The greatest jumpers seem to "hang in the air" in defiance of gravity. Ask your friends to estimate the "hang time" of the great jumpers—the amount of time a jumper is airborne, with feet off the ground. Two or three seconds? Several seconds? Nope. Surprisingly, the hang time of the greatest jumpers is nearly always less than 1 second! The seemingly longer time is one of many illusions we have about nature.

Very few athletes can jump higher than 2 feet. Basketball star Spud Webb won the NBA slam dunk contest in 1986 despite being one of the NBA's shortest players (5 feet 7 inches). Spud, traded from the Sacramento Kings to the Atlanta Hawks in 1995, is probably still the champ, with his vertical jump of 4 feet (1.25 meters). Again, any higher jumps are illusory. It is important to distinguish between how high one can reach and how high one can jump. A very tall basketball player, for example, doesn't have to jump very high to outdistance shorter rivals. Spud, surprisingly, is one of the shorter basketball players, making his jumping ability all the more outstanding. It's important to distinguish between the height of a horizontal bar a jumper clears and the actual vertical displacement of the jumper's center of gravity. Jumpers clearing a high bar can contort their bodies to clear bars that their centers of gravity pass beneath (more about this in Chapter Six).

Jumping ability is best measured by a standing vertical jump. Stand facing a wall and, with feet flat on the floor and arms extended upward, make a mark on the wall at the top of your reach. Then make your jump and at the peak make another mark. The distance between these two marks measures your vertical leap. If its more than 2 feet (0.6 meters), you're exceptional.

Now let's look at the physics. When you leap upward, jumping force is applied only so long as your feet are still in contact with the ground. The greater the force, the greater your launch speed and the higher the jump. It is important to note that as soon as your feet leave the ground, whatever upward speed you attain immediately decreases at the steady rate of g—10 m/s². Maximum height is attained when your upward speed decreases to zero. Then you begin to fall, gaining speed at exactly the same rate, g. Time rising equals time falling: hang time is the sum of time up and time down. While airborne, no amount of leg or arm pumping or other bodily motions can change your hang time.

The relationship between time up or down and vertical height is given by

$$d = \frac{1}{2}gt^2$$

If we know the vertical height, we can rearrange this expression to read

$$t = \sqrt{\frac{2d}{g}}$$

Answers

1. There are several predictions you can make about the dropping cat. To begin with, its acceleration will be 10 m/s², so the speed with which it meets the ground is

- **Speed**: $v = gt = 10\ \text{m}/\text{s}^2 \times \frac{1}{2}\text{s} = 5\ \text{m}/\text{s}$

- Average speed will be: $\bar{v} = \dfrac{\text{initial } v + \text{final } v}{2} = \dfrac{0\ \text{m}/\text{s} + 5\ \text{m}/\text{s}}{2} = 2.5\ \text{m}/\text{s}$

We put a bar over the symbol to denote average speed $\bar{v}$.

- The distance it drops is: $d = \bar{v}t = 2.5\ \text{m}/\text{s} \times \frac{1}{2}\text{s} = 1.25\ \text{m}$

Or equivalently,

$$d = \frac{1}{2}gt^2$$
$$= \frac{1}{2} \times 10\ \text{m}/\text{s}^2 \times (\frac{1}{2}\text{s})^2 = \frac{1}{2} \times 10\ \text{m}/\text{s}^2 \times \frac{1}{4}\text{s}^2 = 1.25\ \text{m}$$

Notice that we can find the distance by either of these equivalent relationships. Doing it both ways is a good check!

2. If air resistance is negligible, acceleration will be the same, 10 m/s², which means its speed will still increase 10 m/s each second. However, if air resistance is present, then as we'll see in the next chapter, more than gravity is acting on the thrown object so acceleration will be less. We'll see that air drag reduces the net force and reduces the acceleration.

Let's use Spud's jumping height of 1.25 meters for d, and use the more precise 9.8 m/s² for g. Solving for t, half the hang time, we get

$$t = \sqrt{\frac{2d}{g}} = \sqrt{\frac{2(1.25)}{9.8}} = 0.5 \text{ second.}$$

Double this (because this is the time for one way of an up-and-down round trip) and we see that Spud's record-breaking hang time is 1 second.*

We've only been talking about vertical motion. How about running jumps? We'll learn in Chapter 8 that hang time depends only on the jumper's vertical speed at launch and is independent of horizontal speed. While airborne the jumper moves horizontally at a constant speed, while only vertical speed undergoes acceleration. Interesting physics!

Velocity Vectors

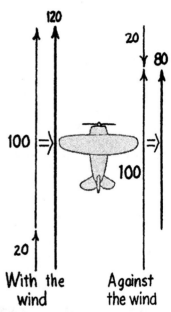

In the previous chapter we treated vectors as they apply to forces. Velocity has both magnitude and direction and is also a vector quantity. The magnitude of velocity is speed. We combine vectors as we did in the previous chapter. When vectors act along parallel directions, they combine by addition. When in opposite directions, they subtract. The sum of two or more vectors is their resultant.

Consider, for example, an airplane flying due north at 100 km/h relative to the surrounding air. Suppose there is a tailwind (wind from behind) that moves due north at 20 km/h. A little thought shows the groundspeed (speed relative to the ground below) is then 120 km/h. With the tailwind, it would travel 120 kilometers in one hour. Now suppose the airplane makes a U-turn and flies south into the wind (wind head-on). The resultant is 100 km/h minus 20 km/h, which equals 80 km/h. Flying against a 20 km/h headwind, the airplane would travel only 80 kilometers relative to the ground in 1 hour.

Now consider a more complicated example. Suppose an airplane flies in a cross-wind. Will the groundspeed be affected? Many students say no, reasoning that when the wind is at 90 degrees it is neither against nor with the wind. But it so happens that an airplane will be blown offcourse and have an

---

* A general rule is that height jumped in feet is equal to four times hang time squared:
$$d = \frac{g}{2}\left(\frac{T}{2}\right)^2 = \frac{g}{2}\left(\frac{T^2}{4}\right) = \frac{g}{8}(T^2) = \frac{32}{8}(T^2) = 4T^2. \text{ (In feet, } g = 32 \text{ ft/s}^2\text{)}$$

increased speed in a crosswind. We can see this with vectors. Consider a slow-moving airplane that flies 80 km/h north and is caught in a strong westerly crosswind of 60 km/h. The sketch shows vectors for the airplane velocity and wind velocity. The scale here is 1 cm: 20 km/h. The diagonal of the constructed parallelogram (rectangle in this case) measures 5 cm. This diagonal represents the resultant velocity, which in this case is 100 km/h. So the airplane moves at 100 km/h relative to the ground, in a direction between north and northeast.* The sketch of the airplane is the way I drew them when I was a kid. The planes I used to make out of folded paper, however, looked nothing like a "real" airplane back then. How ironic that the paper airplanes look so much like the stealth bomber and other modern planes of today!

Question

What is the maximum resultant of a pair of 30-m/s velocities? The minimum resultant?

Here's another Practice Page that illustrates the parallelogram rule. Can you do constructions c and d for Question 1? And can you find the resultants in a-d for Question 2?

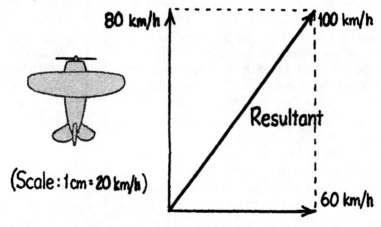

80 km/h   100 km/h

Resultant

(Scale: 1cm = 20 km/h)

60 km/h

Answer

Maximum occurs when the velocities are in the same direction: 60 m/s.
Minimum when they are in opposite directions; 0 m/s.

---

* Whenever the vectors are at right angles to each other, their resultant can be found by the Pythagorean Theorem, a well-known tool of geometry. It states that the square of the hypotenuse of a right-angle triangle is equal to the sum of the squares of the other two sides. Note that two right triangles are present in the parallelogram shown (rectangle in this case). From either one of these triangles we get:

$$\text{resultant}^2 = (60 \text{ km/h})^2 + (80 \text{ km/h})^2$$
$$= (3600 \text{ km}^2/\text{h}^2) + (6400 \text{ km}^2/\text{h}^2)$$
$$= (10,000 \text{ km}^2/\text{h}^2)$$

The square root of $(10,000 \text{ km}^2/\text{h}^2)$ is 100 km/h, as expected.

CONCEPTUAL **Physics** PRACTICE PAGE

Vectors and the Parallelogram Rule

1. When vectors A and B are at an angle to each other, they add to produce the resultant C by the parallelogram rule. Note that C is the diagonal of a parallelogram where A and B are adjacent sides. Resultant C is shown in the first two diagrams, a and b. Construct the resultant C in diagrams c and d. Note that in diagram d you form a rectangle (a special case of a parallelogram).

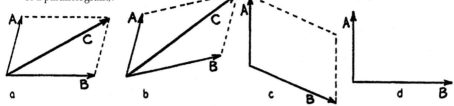

2. Below we see a top view of an airplane being blown offcourse by wind in various directions. Use the parallelogram rule to show the resulting speed and direction of travel for each case. In which case does the airplane travel fastest across the ground? _____ Slowest? _____

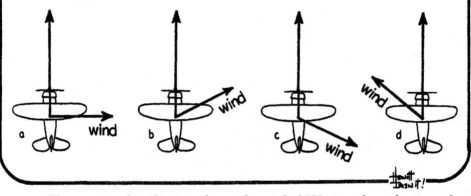

Hewitt-Drew it!

   Please remember that it took people nearly 2000 years from the time of Aristotle to reach a clear understanding of motion, so be patient with yourself if you find that you require a few hours to achieve as much! Likewise with the chapters that follow: it takes time for the ideas of physics to become part of your thinking. Unless learning is accompanied by laboratory investigation, or tied to direct experience, what you learn about physics is consequently limited. Every physics teacher I've known says they didn't experience a full understanding of physics until teaching it—in lecture, problem-solving sessions, and in lab—long after first studying it. So it would be unrealistic to hope you might gain a full understanding in reading this book. What most readers will gain, however, is enough knowledge to elicit the nice feeling that comes with "getting it." Understanding may require more than one reading of some sections, for some catch on fast while others, like myself, need more time. Whatever your pace, the satisfying feeling that comes with learning a deep thing well should be a delightful outcome for most readers, but only with what's required: concentration, effort, and patience.

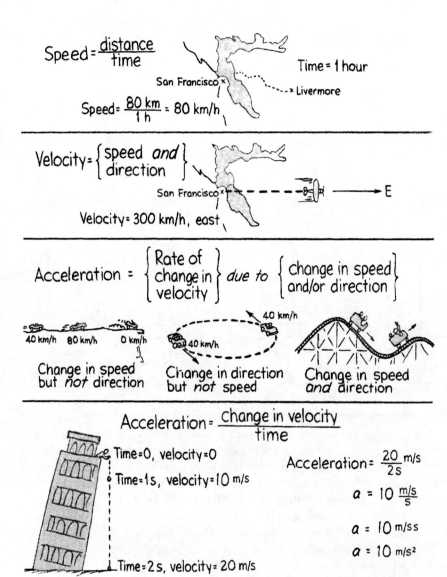

Speed = $\dfrac{\text{distance}}{\text{time}}$

San Francisco ×   ---× Livermore

Time = 1 hour

Speed = $\dfrac{80\ \text{km}}{1\ \text{h}}$ = 80 km/h

Velocity = $\left\{\begin{array}{l}\text{speed } and \\ \text{direction}\end{array}\right\}$

San Francisco ×

→ E

Velocity = 300 km/h, east

Acceleration = $\left\{\begin{array}{l}\text{Rate of} \\ \text{change in} \\ \text{velocity}\end{array}\right\}$ *due to* $\left\{\begin{array}{l}\text{change in speed} \\ \text{and/or direction}\end{array}\right\}$

40 km/h

40 km/h   80 km/h   0 km/h

40 km/h

Change in speed
but *not* direction

Change in direction
but *not* speed

Change in speed
and direction

Acceleration = $\dfrac{\text{change in velocity}}{\text{time}}$

Time=0, velocity=0

Time=1s, velocity=10 m/s

Acceleration = $\dfrac{20}{2\text{s}}$ m/s

$a = 10\ \dfrac{\text{m/s}}{\text{s}}$

$a = 10\ \text{m/ss}$

$a = 10\ \text{m/s}^2$

Time=2s, velocity= 20 m/s

3

Newton's Laws of Motion

Now that we can distinguish between velocity and acceleration, we're ready for the good stuff—Newton's laws of motion, which have been around for three centuries and are still the backbone of classical physics. If you were to begin your study of physics by learning modern concepts such as relativity and quantum mechanics without a knowledge of Newton's laws, your physics knowledge would be severely limited. Newton's laws are the essence of classical physics and provide the foundation for understanding more modern concepts. Whatever the insights of modern physics developed in this century, it's safe to say that Newton's three laws of motion and his law of universal gravitation will remain central concepts of physics in general. The simplicity of Newton's laws makes them not only beautiful, but enormously useful. They are all that is needed to precisely get rockets to the moon and beyond.

The origin of Newton's laws began with Galileo, who died in 1642, within a year of Newton's birth. This was just after Shakespeare died, when John Milton was writing, and when the East Coast of America was being colonized. Newton's famous quote, "If I have seen farther than others, it is because I have stood on the shoulders of giants" certainly must include Galileo, who broke with the teachings of Aristotle that held sway during Galileo's time. Whereas Aristotle claimed that moving objects must be steadily propelled by forces to keep them moving, Galileo showed with his inclined plane experiments that moving things, once moving, continued in motion *without* the application of forces. He called the property of objects to behave this way *inertia*—which comes from the Latin word *iners* ("lazy" or "inert"). The stage was set for Isaac Newton.

By the time Newton was twenty-three, he had developed his famous three laws of motion that completed the overthrow of Aristotelian ideas. These three important laws first appeared in one of the most important books of all time, Newton's *Principia* (*Philosophiae Naturalis Principia Mathematica*).* The first law is a restatement of Galileo's concept of *inertia*; the second law relates acceleration to the cause that produces it, *force*; and the third is the law of *action* and *reaction*. We begin with the first of these laws, the crux of which was the basis of equilibrium as discussed in Chapter One.

Newton's First Law of Motion

The tendency of things to resist changes in motion was what Galileo called *inertia*. Newton refined Galileo's idea and made it his first law, appropriately called the **law of inertia**. From Newton's *Principia*:

> **Law 1: Every material object continues in its state of rest, or of uniform motion in a straight line, unless it is compelled to change that state by forces impressed upon it.**

There are two key words in this law: the first is *continues*. An object *continues* to do whatever it happens to be doing unless a force is exerted upon it. If it is at rest, it *continues* in a state of rest. If it is moving, it *continues* to move without turning or changing its speed. The second key word is *change*: objects don't resist motion—rather, they resist a *change* in their existing motion—they

resist being accelerated. An object does not accelerate of itself; acceleration must be imposed against the tendency of an object to retain its state of motion. Things at rest tend to stay at rest—things moving tend to continue moving. **Inertia** is the property of bodies to resist changes in motion.

We use the principle of inertia when we stamp our feet to remove snow from them, shake a garment to remove dust, and tighten the loose head of a hammer by slamming the hammer handle-side-down on a firm surface. Probably the most celebrated demonstration of inertia is the classic tablecloth-and-dishes stunt. When the tablecloth is yanked properly and suddenly, the dishes remain intact on the table. Another demonstration of inertia consists of two globs of clay stuck to the ends of a coat-hanger wire bent into an M shape. Balance the apparatus on your head. With one blob in front of your face, ask how you can see the other

---

* "The Mathematical Principles of Natural Philosophy."

blob without touching the apparatus. Then simply turn around and there it is! My friend Marshall Ellenstein likens this to the bowl of soup you turn only to find the soup stays put. Inertia in action!

Question

When we jump straight up, why do we land in our footsteps rather than at a location equal to the distance the earth moves during our jump?

An object at rest, as we learned in Chapter One, is in equilibrium. But an object can be in equilibrium when it's moving. A hockey puck is in equilibrium both when it is at rest and when it is sliding at constant velocity across the ice. In the first case, we say it's in *static* equilibrium; in the second case we say it's in *dynamic* equilibrium. When you push a crate at constant speed across a factory floor, the crate is in dynamic equilibrium. In this case the force you apply is balanced by the friction force between the crate and the floor. With zero net force, the motion is steady—acceleration is zero. Note

that zero acceleration doesn't mean zero speed. Zero acceleration means no *change*—that the object will maintain the speed it happens to have, neither speeding up, slowing down, or changing direction.

Why is the First Law of Motion not broken into two laws: one for objects at rest, and another for objects in uniform motion? Newton showed that an object at rest was just a special case of an object moving at constant velocity (which opened a giant crack in Aristotle's assertion that objects at rest had a special status). At rest, an object's velocity is zero, merely one of an infinite number of possible velocities. Zero velocity has no more significance than non-zero velocities in the first law. In the dining car of a train, for example, the tablecloth-and-dishes stunt yields the same result whether the train has a constant velocity of zero or a constant velocity of 100 km/h. Likewise, if the ride is smooth, a waiter pours coffee into your cup in the same way as if the train were at rest.

Recall from the previous chapter that when we cite the motion of something, we mean relative to something else—usually the earth. The earth serves as a *frame of reference*. Newton's first law relates to frames of reference. We see that a frame of reference at rest and a frame of reference moving at constant

---
Answer

We return to our footsteps because we are already moving along with the moving earth, and in accord with Newton's first law, we remain in that motion during the jump.

---

velocity are not all that different. A ball tossed back and forth behaves the same in a train at rest and in a train moving at constant velocity. In both frames, Newton's first law holds equally. A frame of reference in which Newton's first law holds is called an *inertial* frame of reference. We'll see later that in an accelerating frame of reference, such as a rotating merry-go-round, tossing a ball back and forth or pouring coffee is very different than doing the same when the merry-go-round is at rest. An accelerating reference frame is a non-inertial reference frame. Newton's first law doesn't hold in a non-inertial reference frame. Interestingly, and strictly speaking, the earth is a non-inertial frame because of its rotation. However, unless we consider large-scale motions such as wind and ocean currents, we can usually make the approximation that the earth is an inertial frame. That approximation is made in the rest of this book.

Mass

Every material object possesses inertia: how much depends on its amount of matter—the more matter, the more inertia. In speaking of how much matter something has, we use the term *mass*. The greater the mass of an object, the greater the amount of matter and the greater its inertia. **Mass** is a measure of the inertia of a material object.

Loosely speaking, mass corresponds to our intuitive notion of **weight**. We say something has a lot of matter if it's heavy, because we're accustomed to measuring matter by gravitational attraction to the earth. But mass is more fundamental than weight; it is a fundamental quantity that completely escapes the notice of most people. There are times, however, when weight corresponds to our unconscious notion of inertia. For example, if we are trying to determine which of two small objects is the heavier, we might shake them back and forth in our hands or move them in some way instead of lifting them. In doing so, we are judging which of the two is more difficult to accelerate, seeing which is the more resistant to a change in motion. We are really making a comparison of the inertias of the objects.

Mass and weight are confused mainly because they are directly proportional to each other. If the mass of an object is doubled, its weight is also doubled; if the mass is halved, its weight is halved. But there is a distinction between the two.

> **Mass is the quantity of matter in a material object. More specifically, it is the measure of the inertia or sluggishness that an object exhibits in response to any effort made to start it, stop it, or change in any way its state of motion.**

> **Weight is the gravitational force exerted on an object by the nearest most-massive body (locally, by the earth).**

In the United States, we measure the amount of matter in an object by the gravitational pull between it and the earth—its weight, commonly expressed in *pounds*. In most of the world, however, the measure of matter is commonly expressed in a mass unit, the **kilogram**. At the surface of the earth, a brick with a mass of 1 kilogram weighs 2.2 pounds. In the metric system of units, the unit of force is the **newton**, which is equal to a little less than a quarter pound (like the weight of a quarter-pounder hamburger *after* it is cooked). A brick with a mass of 1 kilogram weighs about 10 newtons (10 N).* Away from the earth's surface, where the influence of gravity is less, the same brick weighs less. It would also weigh less on the surface of planets with less gravity than the earth. On the moon's surface, for example, the gravitational force on things is only 1/6 as strong as on earth, so a 1-kilogram object weighs about 1.6 newtons (or 0.4 pounds). On planets with stronger gravity it would weigh more. *But the mass of the brick is the same everywhere.* The brick offers the same resistance to speeding up or slowing down regardless of whether the earth, moon, or whatever is attracting it. In a drifting space ship where a scale with a brick on it reads zero, the brick still has mass. Even though it doesn't press down on the scale, the brick has the same resistance to a change in motion it has on earth. Just as much force would have to be exerted by an astronaut in the space ship to shake the brick back and forth as would be required to shake it back and forth while on earth. It would take the same push to accelerate a Cadillac limousine up to 60 miles per hour on a level surface on the moon as on earth. The difficulty of lifting it against gravity (weight) is something else. Mass and weight are different from each other.

A clever classroom demonstration that distinguishes between mass and weight is a large ball suspended by a string, as shown in the sketch. A second piece of string is attached to the lower part of the hanging ball, which when slowly pulled causes the top string to break. After the broken string is replaced and the pulling is repeated, a quick jerk breaks the bottom string. The instructor asks which of these cases illustrates the role of weight, and which case illustrates the role of inertia. When the bottom string is gradually pulled downward, tension builds up in both strings. But even before the pulling, the tension in the top string is already appreciable because of the ball's weight. Tension will always be greater in the top string, which is why it finally breaks. So, the role of weight results in the top string breaking. But when jerked, the tendency of the ball to resist the sudden downward acceleration, its inertia, results in the lower string break-

---

* So 2.2 lb equal 10 N (more precisely 9.8 N), or 1 N is approximately equal to 0.22 lb—about the weight of an apple. In the metric system it is customary to specify quantities of matter in units of mass (in grams or kilograms) and rarely in units of weight (in newtons). In the United States, however, quantities of matter are customarily specified in units of weight (in pounds), and the unit of mass, the *slug*, is not well known.

I teach that weight = *mg*, the gravitational force on a body. Some physicists prefer to define weight as the force a body exerts on a surface (or on a support, if suspended). More about this in Chapter 7.

ing. The different effects of weight and mass are nicely shown.

A dramatic follow-up demonstration is lying on your back and having an assistant place a blacksmith's anvil on your stomach and then strike it rather hard with a sledge hammer. The principles here are the same as the ball and string demo. Both the inertia of the ball and the inertia of the anvil resist the changes in motion they would otherwise undergo. Just as the string doesn't break, your body is not squashed. (Be sure your assistant is competent with a hammer. In my fourth year of teaching, the student who volunteered was extra nervous in front of the class and missed the anvil entirely—but not me. The hammer smashed into my hand breaking two fingers. I was lucky not to be harmed more!)

Question

Ask a friend to drive a small nail into a piece of wood placed on top of a pile of books on your head. Why doesn't this hurt you?

Some people confuse mass and **volume**. A massive object doesn't always mean a big object. Which is easier to get moving: a car battery or an empty cardboard box of the same size? So, we see that mass is neither weight nor volume.

Answer

The relatively large mass of the books and block atop your head resist being moved. The force that is successful in driving the nail will not be as successful in accelerating the massive books and block, which don't move very much when the nail is struck.

Can you see the similarity of this to the suspended massive ball demonstration, where the supporting string doesn't break when the bottom string is jerked?

Questions

1. A hockey puck sliding across the ice finally comes to rest. How would Aristotle interpret this behavior? How would Galileo and Newton interpret it? How would you interpret it?

2. Does a 2-kg iron brick have twice as much *inertia* as a 1-kg iron brick? Twice as much *mass*? Twice as much *volume*? Twice as much *weight*?

3. Would it be easier to lift a large truck on the earth or to lift it on the moon?

When we compare mass to volume, or weight to volume, we're talking about *density*. Something with a lot of mass in a small volume, like a gold nugget, has a high density. Something with a small mass in a relatively large volume, like Styrofoam, has a low density. Density is neither mass nor volume; density is the *ratio* of mass/volume. (More about density when we discuss materials in Book Two).

To summarize, although Galileo introduced the idea of inertia, Newton grasped its significance. The law of inertia defines natural motion and tells us what kinds of motion are the result of applied forces. Whereas Aristotle maintained that the forward motion of an arrow through the air required a steady force, Newton's law of inertia instead tells us that the behavior of the arrow is natural; constant speed along a straight line (or, simply, constant velocity) requires no force. Aristotle and his followers held that the circular motions of

Answers

1. Aristotle would probably say that the puck slides to a stop because it seeks its proper and natural state, that of rest. Galileo and Newton would probably say that once in motion the puck would continue in motion, and that what prevents continued motion is not its nature or its proper rest state, but the friction between the puck and the ice. This friction is small compared to the friction between the puck and a wooden floor, which is why the puck slides so much farther on ice. Only you can answer the last question.

2. The answers to all parts are yes. A 2-kg iron brick has twice as many iron atoms and therefore twice the amount of matter and mass. In the same location, it also has twice the weight. And since both bricks have the same density (the same mass/volume), the 2-kg brick also has twice the volume.

3. A truck would be easier to lift on the moon because gravity is less on the moon. When you lift an object, you are contending with the force of gravity (its weight). Its weight is only 1/6 as much on the moon, so only 1/6 as much effort is required to lift it there. To move it horizontally, however, you are not pushing against gravity. Mass is the only factor, and is the same on the earth or the moon (or anywhere), so equal forces will produce equal accelerations whether the object is on the earth or the moon.

heavenly bodies were natural and required no applied forces. The law of inertia, however, clearly states that in the absence of forces of some kind the planets would not move in the divine circles of ancient and medieval astronomy but would move instead in straight-line paths off into space. Newton maintained that the curved motion of the planets was evidence of some kind of force. We shall see in Chapter Seven that his search for this force led to the law of gravity.

Net Force

Newton's first law tells us that when an object is at rest, or moving at constant velocity, then no force acts on the object. More specifically, no *net* force acts on the object. A force, in the simplest sense, is a push or a pull. Its source may be gravitational, electrical, magnetic, or simply muscular effort (which we'll learn later is actually electrical in nature). We say *net* force because often more than a single force acts on an object. If, for example, you and a friend pull in the same direction with equal forces on an object, the forces add to produce a net force twice as great as your single force. We'll soon see that the combination of forces will produce twice the acceleration than if you

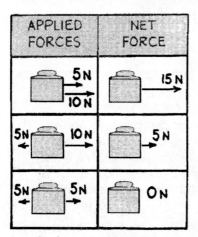

pulled alone. If, however, your equal pulls are in opposite directions, no acceleration occurs. Because they are oppositely directed, the forces on the object cancel each other. One of the forces can be considered to be the negative of the other, and they add algebraically to zero. So net force = 0. This is the equilibrium rule of Chapter One: $\Sigma F = 0$.

Recall from Chapter One that a quantity that involves both magnitude (the amount) and direction is classified as a vector quantity. Force is a *vector quantity*. When two or more forces are exerted on an object, the forces combine by vector rules. The vector rule for combining parallel forces, in equal or opposite directions as just discussed, is simple algebraic addition. When two or more forces are exerted at angles to one another, the parallelogram rule is used to find the net force—the *resultant*.

Friction

Most things that are moved in our environment must be pushed or pulled to overcome friction. The direction of friction force is always in a direction opposing motion. Thus, if an object is to move at constant velocity, a force equal to the opposing force of friction must be applied so that the two forces exactly cancel each other. The zero net force then results in zero acceleration.

Friction occurs when surfaces slide or tend to slide over one another. The amount of friction depends on the kinds of material, the mutual contact of irregularities in the surfaces, and how much they are pressed together.* The irregularities act as obstructions to motion. Even surfaces that appear to be very smooth, when viewed through a microscope, can be seen to have irregular surfaces.

It is said that wisdom is knowing what to overlook. Here we see the wisdom of Galileo and Newton evident as they looked beyond friction in their search for the laws of motion. Their success in uncovering nature's laws had much to do with their ability to distinguish what is central and what is peripheral about a situation—to separate first order effects that are basic from second, third, or even fourth-or-more order effects. They looked beyond the effects of air resistance, other frictional forces, and other things such as spin and the shapes of moving objects. In so doing they found simple relationships that these factors had obscured for Aristotle.

Question

A jumbo jet cruises at constant velocity of 600 mi/h when the thrusting force of its engines is a constant 800,000 pounds. What is the acceleration of the jet? What is the force of air resistance on the jet?

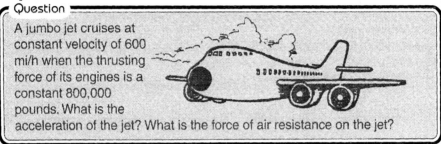

Newton's Second Law of Motion

In the previous chapter we discussed motion without regard to forces—*kinematics*. Here we study motion and the forces that contribute to changes in motion—*dynamics*. Most of the motion we observe undergoes changes—accelerated motion, which is the result of one or more applied forces. The overall net force, whether it be from a single source or a combination of sources,

Answer

The acceleration is zero because no change in speed or direction occurs. Since the acceleration is zero, the net force is zero, which in turn means that the force of air resistance must just equal the thrusting force of 800,000 pounds and act in the opposite direction. So the air resistance on the jet is 800,000 pounds. (Note that we don't need to know the speed of the jet to answer this question. We need only to know that it is constant, our clue that acceleration is zero, and, therefore, net force is zero.)

---

* Even though it may not seem so yet, most of the concepts in physics are not really complicated. But friction is different—it is a very complicated phenomenon. The findings are empirical (gained from a wide range of experiments) and the predictions are approximations (also based on experiment).

produces acceleration. The relationship of acceleration to force and inertia is given in Newton's second law.

Law 2: The acceleration of an object is directly proportional to the net force acting on the object, is in the direction of the net force, and is inversely proportional to the mass of the object.

In summarized form, this is

$$\text{Acceleration} \sim \frac{\text{net force}}{\text{mass}}$$

In symbol notation, this is simply

$$a \sim \frac{F_{net}}{m}$$

We use the wiggly line ~ as a symbol meaning "is proportional to." (Mathematicians and many physics instructors prefer the symbol $\propto$.) Either way, we say that acceleration a is directly proportional to the overall net force F and inversely proportional to the mass m. By this we mean that if F increases, a increases by the same factor (if F doubles, a doubles); but if m increases, a decreases by the same factor (if m doubles, a is cut in half). With appropriate units of F, m, and a, the proportionality may be expressed as an exact equation:*

$$a = \frac{F_{net}}{m}$$

Force of hand accelerates the brick

An object is accelerated in the direction of the force acting on it. Applied in the direction of the object's motion, a force will increase the object's speed. Applied in the opposite direction, it will decrease the speed. Applied at right angles, it will deflect the object. Any other direction of application will result in a combination of speed change and deflection. *The acceleration of an object is always in the direction of the net force.*

Twice as much force produces twice as much acceleration

In the second law, Newton gives a more precise idea of force by relating it to the acceleration it produces. He says in effect that *force is anything that can accelerate an object.* The larger the force, the more acceleration it produces. Acceleration is directly proportional to net force.

Twice the force on twice the mass gives the same acceleration

* This equation is most often written, $F = ma$, which is algebraically equivalent, but which to my way of thinking conveys a wrong message. It implies that force is mass times acceleration. A force, however, is a push or a pull, or as we'll see when we treat Newton's third law, part of an interaction between one body and another.

Questions

1. Acceleration was defined in Chapter Two as the time rate of change of velocity; that is, a = (change in v)/$time$. Are we in this chapter saying that acceleration is instead the ratio of force to mass; that is, $a = F/m$? Which is it?

2. How does Newton's second law, $a = F/m$, help explain why a rocket becomes progressively easier to accelerate as it travels through space?

Force and mass have opposite effects on acceleration. The more massive the object, the less its acceleration. For the same force, twice the mass results in half the acceleration; three times the mass, one-third the acceleration. So we see that increasing the mass decreases the acceleration. For example, if we put identical Ford engines in a Peterbilt truck and a Honda Civic, we would expect quite different accelerations even though the driving force in each car is the same. The Peterbilt truck with its greater mass has a greater resistance to a change in velocity than does the Honda Civic. Consequently, the more massive truck requires a more powerful engine to achieve the same acceleration. For the same acceleration, a larger mass requires a correspondingly larger force. We say that acceleration is inversely proportional to mass.*

Newton's second law is extremely useful in analyzing motion. Measure the mass of an object, impress a given force on it, and we can predict the acceleration. This law, like most laws, is only an *approximation* of what occurs in nature. Although the mass of an

Force of hand accelerates the brick

The same force accelerates 2 bricks ½ as much

3 bricks, ⅓ as much acceleration

Answers

1. Chapter Two tells us what acceleration is: this chapter tells us how it is produced. Acceleration is defined as the time rate of change of velocity and is *produced* by a force. How much force/mass (the cause) determines the velocity change per unit of time (the effect). So whereas we defined acceleration in Chapter Two, in this chapter we define the terms that cause acceleration.

2. As fuel is burned, the mass of the rocket m decreases. So a increases! There is simply less mass to be accelerated as fuel is consumed.

---

* Mass can be operationally defined as the proportionality constant between force and acceleration in Newton's second law, rearranged to read $m = F/a$. A 1-unit mass is that which requires 1 unit of force to produce 1 unit of acceleration. So 1 kg is the amount of matter that 1 N of force will accelerate 1 m/s². (In British units, 1 slug is the amount of matter that 1 pound of force will accelerate 1 ft/s².) We shall see later that mass is simply a form of concentrated energy.

object appears to be unaffected by motion, we will see in Book Five that increases in the momentum of fast-moving particles are greater than comparable increases in speed, which in effect means that mass increases with speed. This increase is not appreciable unless the speed is near the speed of light. For speeds less than 100 miles per second, for example, the mass is constant to within one part in a million. So for ordinary speeds, we can say acceleration is very nearly equal to the impressed force divided by the mass. This approximation won't get us in trouble. With it we can put space probes very accurately on distant planets. The fact that the law is approximate, however, is interesting philosophically in that it suggests a deeper law of which the second law is a special case (we'll treat that deeper law when we investigate special relativity in Book Five). Even a small effect sometimes leads to a profound change in our ideas about nature. One of the beauties of physics is that exceptions to a law often point the way to a more fundamental law. New laws are not discovered by dreamers, as a lot of people suppose, but by those acquainted with existing laws well enough to know where the approximations are and where potential cracks can be located. Any "good luck" in scientific investigation goes to those who are prepared to see what others have missed.

Applying Force—Pressure

Put a book on a table and no matter how its placed—on its back, upright, or even balanced on a single corner—the force of the book against the table is the same. You can check this by placing a book in different orientations on a bathroom scale where you'll read the same weight in all cases. But do the same on the palm of your hand and the same weight feels different. That's because the area of contact will be different for different orientations, producing different pressures. The force per unit area is called **pressure**.

$$\text{Pressure} = \frac{\text{force}}{\text{area of application}}$$

where the force is perpendicular to the surface area. In equation form:

$$P = \frac{F}{A}$$

where P is the pressure and A is the area against which the force is applied. Force, measured in newtons, is different from pressure, which is measured in newtons per square meter.

Stand on a floor and you exert pressure against the floor. Stand on one foot and you double the pressure because the area of contact is half. Stand on a spike heel, and the pressure you exert may permanently dent the floor (especially a linoleum surface). The smaller the area that supports a given force, the greater the pressure on that surface. The books on the table both exert the same force against the table—but different pressures. The upright book exerts appreciably more pressure on the table.

Which provides more friction between a car tire and the road—a wide tire or a narrow tire? Most people will say a wide tire and point to the wide tires on racing vehicles. But like the book on the table, the *force* against the surface is the same. It turns out that the *force* of friction is proportional to this force and not the pressure. So why the wider tires? To spread the force out and reduce wear. A narrow tire with its greater pressure would heat up more quickly and wear away faster.

A most dramatic illustration of pressure is a classroom demo that Pablo Robinson and I have done many times over the years in our classes. Pablo lies between two beds of sharp nails, atop which is placed a cement block. Students scream as I smash the block to bits with a sledgehammer. Why isn't Pablo hurt? Fortunately, about 200 nails make contact with his body. The combined surface area of this many nails results in a pressure that does not puncture the skin. Much area, little pressure. So Pablo, somewhat welted, is spared to repeat the demonstration another time. He doesn't always go unscathed, however. When the demonstration is repeated for more than one class on the same day, the welts from one encounter often line up with the nails during a successive encounter. Ouch! When this happens, Pablo does bleed a bit. There are dedicated teachers, and not-so-dedicated teachers. Pablo has demonstrated many times that he is among the first category. (He also was the California Presidential Physics Teacher of the year for 1988.)

Another dedicated teacher, dear friend, Marshall Ellenstein, demonstrates pressure by walking barefoot over broken glass in his Chicago classroom. Broken glass from smashed bottles is spread over a piece of canvas on the classroom floor. Dramatically, Marshall walks across it unharmed. Like Pablo's nails, there are many points of contact—more area, less pressure. Classroom antics like those of Pablo and Marshall are remembered long after all the formulas, facts, and diagrams have been forgotten. And if the physics underlying them is also forgotten, the all-important savory flavor of physics remains! Flavor is important.

Free Fall—When Acceleration is g

Although Galileo founded both the concepts of inertia and acceleration, and was the first to measure the acceleration of falling objects, he couldn't explain *why* objects of various masses fall with equal accelerations when air drag is

> **Questions**
> 1. In doing the bed of nails demo, would it be wise to begin with a few nails, and work upward to more nails?
> 2. How does the mass of the block play a role in this demonstration?
> 3. If you apply pressure to a solid or liquid object, do you think its temperature will rise as a result of the increased pressure?

negligible. He reasoned that falling bodies should have the same acceleration by a simple argument, which went something like this: if a couple of identical cannonballs were dropped from a tower, we'd expect them to have the same acceleration. Now suppose that in their fall, they were somehow taped together. In effect, they would comprise a falling object with twice the mass. We wouldn't expect the taped-together balls to pick up more speed as a result. It therefore seemed clear to Galileo that the acceleration of freely falling objects doesn't depend on mass. An explanation of why this is so awaited Newton's second law.

We know that a falling object accelerates toward the earth due to the gravitational attraction between the object and the earth. We call this force the *weight* of the object.* When this is the only force—that is, when friction such as air drag is negligible—we say the object is in a state of **free fall**.

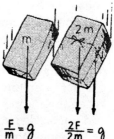

The attractive force of gravity between a more massive object and the earth is greater than that of a less massive object. The double brick shown in the sketch, for

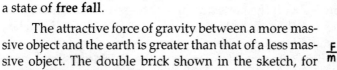

$$\frac{F}{m} = g \qquad \frac{2F}{2m} = g$$

> **Answers**
> 1. No, no, no! Less nails means greater pressure—what you don't want.
> 2. Like the sledgehammer and anvil example, the greater the mass of the block, the smaller its acceleration toward Pablo. It is also important that the block break upon impact (energy of the hammer is best spent in flying cement fragments than in downward motion of the whole block).
> 3. No. Applying pressure to incompressible objects, like most solids and liquids, doesn't increase their temperature. Squeezing a coin in a vice, for example, doesn't warm the coin. And the ocean remains freezing cold at its bottom where pressure is enormous. Many people think that pressure within the earth is responsible for the earth's molten interior. Not so. We'll learn in later books that this heat is due to the energy of radioactivity. Energy must be added to increase the temperature of something. Now if you compress a gas and reduce its volume, the gas warms. We'll learn in thermodynamics that energy is added when volume is compressed. But simple squeezing without a change in volume doesn't increase temperature.

---

* Weight and mass are directly proportional to each other, and the constant of proportionality is *g*. We see that weight = *mg*, so 9.8 N = (1 kg)(9.8 m/s²).

example, is acted upon with twice the attractive force as the single brick. Why then, as Aristotle supposed, doesn't the double brick fall twice as fast? The answer is that the acceleration of an object depends not only on the force—in this case, the weight—but on the object's resistance to motion, its *inertia*. Whereas a force produces an acceleration, inertia is a *resistance* to acceleration. So twice the force exerted on twice the inertia produces the same acceleration as half the force exerted on half the inertia. Both accelerate equally. The acceleration due to gravity is symbolized by g. We use the symbol g, rather than a, to denote that acceleration is due to gravity alone.

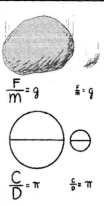

$$\frac{F}{m} = g \qquad \frac{f}{m} = g$$

$$\frac{C}{D} = \pi \qquad \frac{c}{p} = \pi$$

> **Question**
>
> In a vacuum, a coin and a feather fall at equal speeds, side by side. Would it be correct to say that in a vacuum equal forces of gravity act on both the coin and the feather?

    The ratio of weight to mass for freely falling objects equals a constant—g. This is similar to the constant ratio of circumference to diameter for circles—π. The ratio of weight to mass is the same for both heavy and light objects, similar to the way the ratio of circumference to diameter is the same for both large and small circles.

    We now understand that the acceleration of free fall is independent of an object's mass. A boulder 100 times more massive than a pebble falls at the same acceleration as the pebble because although the force on the boulder (its weight) is 100 times greater than the force (or weight)

> **Answer**
>
> No, no, no, a thousand times no! These objects accelerate equally not because the forces of gravity on them are equal (they aren't!), but because the *ratios of their weights to masses* are equal. Although air resistance is not present in a vacuum, gravity definitely acts in a vacuum (you'd know this if you stuck your hand into a vacuum chamber and a Peterbilt truck rolled over it). More gravity force on the coin (weight) is offset by more inertia of the coin (mass). If you answered yes to this question, let this warn you to be more careful when you think physics!

on the pebble, its resistance to a change in motion (mass) is 100 times that of the pebble. The greater force offsets the equally greater mass.

Non-Free fall—When Acceleration is Less Than g

Objects falling in a vacuum or without air drag are one thing, but what of the practical cases of objects falling in air? Although a feather and a coin will fall equally fast in a vacuum, they fall quite differently in air. How do Newton's laws apply to objects falling in air? The answer is that Newton's laws apply to *all* objects, whether freely falling or falling in the presence of resistive forces.

The accelerations, however, are quite different for the two cases. The important thing to keep in mind is the idea of *net force*. In a vacuum or in cases where air drag can be neglected (as in Backyard Physics), the net force is the weight because it is the only force. In the presence of air drag, however, the net force is less than the weight—it is the weight minus air drag.*

The force of air drag experienced by a falling object depends on two things. First, it depends on the frontal area of the falling object—that is, the amount of air the object must plow through as it falls. Second, it depends on the speed of the falling object: the greater the speed, the greater the number of air molecules an object encounters per second and the greater the force of molecular impact. Air drag depends on the size and the speed of a falling object.

In some cases air drag greatly affects falling, in other cases it doesn't. Air drag is important for a falling feather. Since a feather has so much area compared to its small weight, it doesn't have to fall very fast before the upward-acting air drag cancels the downward-acting weight. The net force on the feather is then zero and acceleration terminates. When acceleration terminates, we say the object has reached its **terminal speed**. If we are concerned with direction (down for falling objects), we say the object has reached its **terminal velocity**.

The same idea applies to all objects falling in air. Consider skydiving. As a falling skydiver gains speed, air drag may finally build up until it equals the weight of the skydiver. If and when this happens, the net force becomes zero and the skydiver no longer accelerates; he or she has reached terminal velocity. For a feather, terminal velocity is a few centimeters per second, whereas for a skydiver it is about 200 kilometers per hour. A skydiver may vary this speed by varying position. Head or feet first is a way of encountering less air and thus less air drag and attaining maximum terminal velocity. A smaller terminal velocity is attained by spreading oneself out like a flying squirrel. Minimum terminal velocity is attained when the parachute is opened.

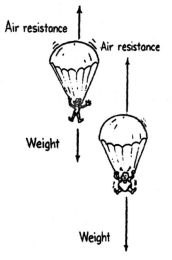

Consider a man and woman parachuting together from the same altitude. Suppose that the man is twice as heavy as the woman and that their same-sized chutes are initially opened. The same size chute means that at equal speeds the air resistance is the same on each. Who gets to the ground first—the heavy man or the lighter woman?

---

* In mathematical notation $a = \dfrac{F_{net}}{m} = \dfrac{mg - R}{m}$ where mg is the weight, and R is the air resistance. Note that when R builds up to mg, then $a = 0$. With no acceleration, the object falls at constant velocity.

The answer is the person who falls fastest gets to the ground first—that is, the person with the greatest terminal speed. At first we might think that because the chutes are the same, the terminal speeds for each would be the same, and, therefore, both would reach the ground at the same time. This doesn't happen because air drag also depends on speed. The woman will reach her terminal speed when air drag against her chute equals her weight. When this happens, the air drag against the chute of the man will not yet equal his weight. He must fall farther and faster than she does for air drag to match his greater weight.* Terminal velocity is greater for the heavier person, with the result that the heavier person reaches the ground first.

Aristotle correctly asserted that heavy things normally fall faster than light things. Where he erred was in attributing this behavior to the weights *and* air drag encountered. It is interesting that today's students often attribute more than is warranted to friction, especially in the laboratory, while the role of friction on motion escaped the notice of Aristotle altogether.

Questions

1. Lillian Lee jumps from a high-flying helicopter. As she falls faster and faster through the air, does her acceleration increase, decrease, or remain the same?
2. How does the adage "You can't change only one thing" apply to the falling skydiver?

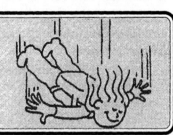

Let's check understanding of the foregoing by considering a pair of tennis balls, one hollow, and the other filled with iron pellets. The iron-filled ball is considerably heavier than the same-size regular ball. If you hold them above your head and drop them simultaneously you'll see that they strike the ground at about the same time. But if you drop them from a greater height, say from the top of a building, you'll note the heavier ball strikes the ground first. Why? In the first case the balls don't gain much speed in their short fall. Air drag encountered is small compared to their weights, even for the regular ball. The tiny difference in their arrival time is likely unnoticed. But when dropped from a greater height, the greater speeds of fall are met with greater air drag. At any given speed each ball encounters the same air drag because each has the same size. This same air drag may be a lot compared to the weight of the lighter ball, but only a little compared to the weight of the heavier ball (like the parachutists discussed on the previous page). For example, 1 N of air drag acting on a 2-N object will cut the acceleration to $g/2$, but 1 N of air drag on a 200-N object will only slightly diminish acceleration. Can you see that even with equal air drags, the accelerations of each are different? There is a

---

* Terminal speed for the man will be about 41 percent greater than terminal speed for the half-as-heavy woman. This is because the retarding force of air resistance is proportional to speed squared. So (man's speed)^2/(woman's speed)^2 = $(1.41)^2$ = 2.

moral to be learned here. Whenever you consider the acceleration of something, use the equation of Newton's second law to guide your thinking: the acceleration is equal to the ratio of *net* force to mass. For the falling tennis balls, the net force on the hollow ball is appreciably reduced as air drag builds up, while comparably the net force on the iron-filled ball is only slightly reduced. Acceleration decreases as net force decreases, which in turn decreases as air drag increases. If and when the air drag builds up to equal the weight of the falling object, the net force then becomes zero and acceleration terminates. Check this out with the equation in the answer to Question #1 on the previous page. Note how all the foregoing is specified in the equation. Another plus for the equations of physics, the value of which is in clearly showing what affects what, and how.

In discussing the effects of air drag on falling objects, it's useful to exaggerate the circumstance so that the effects are more clearly visualized. For example, in comparing the falls of a heavy and light skydiver, think of the lighter person as a feather, and the heavier person as a heavy rock. It's easy to see that air drag plays a more significant role for the falling feather than for the rock. Similarly, but not as much, for the two skydivers.

My friend George Curtis, whose hobby is flying airplanes and who is concerned about clouds and fog, informed me about the role of terminal velocity for airborne clouds and fog. Particles of any mass, in a cloud or otherwise, are pulled downward by earth gravity. The downward pull is greater than the pull on same volumes of surrounding air. So why aren't clouds pulled to earth? The answer is that they are. What counteracts this fall are updrafts of air. Without updrafts, clouds would fall to the earth's surface. Terminal velocity is small for tiny drops and greater for larger drops. When the terminal velocity of the particles is less than the updraft velocity, then those par-

Answers

1. Acceleration decreases because the net force on her decreases. Net force is equal to her weight minus her air drag and, since air drag increases with increasing speed, net force and hence acceleration decrease. By Newton's second law,

$$a = \frac{F_{net}}{m} = \frac{mg - R}{m}$$

where *mg* is her weight and *R* is the air drag she encounters. As *R* increases, *a* decreases. Note that if she falls fast enough so that *R* = *mg*, *a* = 0, then with no acceleration she falls at constant speed.

2. Change any variable on one side of any equation and something on the other side of the equation changes; in this case a change in air drag produces a change in acceleration. This adage has universal application.

ticles are kept from falling to earth. Small drops attain terminal velocity within meters of fall. A reasonable cumulus cloud has updrafts of at least 1 m/s. Fog drops have very low terminal velocity (on the order of cm/s) and fall very slowly. At 1 cm/s it take 100 seconds to fall 1 meter. Typical rain drops of 2 millimeters in diameter have a terminal velocity of about 6.5 m/s. If it weren't for air drag, it would be dangerous to be outdoors on a rainy day!

Question

Why is it that a cat that falls from the top of a 50-story building will hit the ground no faster than if it fell from the 20th story?

Newton's Third Law of Motion

Drop a sheet of tissue paper in front of the heavyweight boxing champion of the world and challenge him to hit it in midair with 50 pounds (222 N) of force. Sorry, the champ can't do it. In fact, his best punch couldn't even come close. Why is this?

So far we've treated force in its simplest sense—as a push or pull. In a broader sense, a force isn't a thing in itself but makes up an *interaction* between one thing and another. Punch a heavy bag and it dents. Pull on a cart and it accelerates. Whack a stake with a hammer and the stake is driven into the ground. One object interacts with another object. Which exerts the force and which receives the force? Newton's answer was that neither force has to be identified as "exerter" or "receiver": both constitute a single interaction. For example, when we hit the punching bag, the bag presses back on us; pull the cart, the cart likewise pulls on us, as evidenced perhaps by the tightening of the rope wrapped around our hand. This pair of forces, our pull on the cart and the cart's pull on us, make up the single interaction between us and the cart. In the interaction between the hammer and the stake, the hammer exerts a force against the stake, but is itself brought to a halt in the process.

Answer

The cat reaches terminal velocity in a 20-story fall, so falling the extra distance doesn't affect speed. The low terminal velocities of small creatures enables them to fall without harm from heights that would kill larger creatures. Falling is favored by their higher ratio of surface area/mass. Humans boost this ratio by using parachutes.

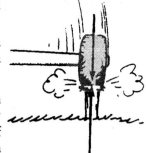

Such observations led Newton to his third law—the law of action and reaction.

Law 3: Whenever one object exerts a force on a second object, the second object exerts an equal and opposite force on the first.

Newton's third law is often stated, "To every action there is always an equal opposed reaction." In any interaction there is an action and reaction pair of forces that are equal in magnitude and opposite in direction. Neither force exists without the other. Forces come in pairs, one action and the other reaction, which make up one interaction between two things. Since a minimum of two objects are needed for an interaction, Newton's third law is sometimes called the *law of interaction*. Getting back to the champ punching the tissue paper: whereas a 50-pound interaction between the champ's fist and an opponent's jaw is a common occurrence, in the swiftest of punches, a 50-pound interaction between his fist and a sheet of tissue paper in midair is not possible. No can do!*

> **Question**
>
> A car accelerates along a road. Exactly what is the force that moves the car?

A teacher in a classroom soon convinces his or her students that the pair of action-reaction forces are equal in magnitude when students are challenged to pull on opposite ends of a rope with unequal forces. A spring scale on each end shows that in no way can different readings simultaneously occur on the scales however the ends are pulled. Actually there are two force pairs here: one consisting of the hand pulling the left side of the rope (hand on rope; rope on hand) and the other force pair at the right end of the rope (rope on hand; hand on rope). A single tension is produced all along the rope, which is read by spring scales anywhere along the rope. The readings will be identical

> **Answer**
>
> The *road* provides the force to accelerate the car! Really! The tire pushes back on the road, as evidenced by dust thrown back when the road is dusty, and the road simultaneously pushes forward on the tire. The tire is connected to the car, so the car accelerates. Now you can look at roadways in a new way!

---

* Even if the mass of the paper were 1 gram, and brought up to a punch speed of 90 miles per hour (40 m/s) in 0.01 seconds, the force required would be only 0.4 N, or 0.09 pounds. A force of 50 pounds exerted for 0.01 second would accelerate the paper to some 2200 m/s, more than six times the speed of sound!

whether the rope is at rest or accelerating (assuming the rope's mass is negligible).

Who wins in a tug of war—the team that pulls harder on the rope, or the team that pushes harder against the floor? Rope tension will be the same for both ends of a rope, so the greater net force is produced by the team that pushes harder against the floor. This is nicely demonstrated on waxed floors when a team of girls with rubber-soled shoes pulls against a team of boys wearing socks. The girls always win!

You interact with the floor when you walk on it. Your push against the floor is coupled to the floor's push against you. The pair of forces occur simultaneously. Likewise, the tires of a car push against the road while the road pushes back on the tires—the tires and road push against each other. In swimming you interact with the water that you push backward, while the water pushes you forward—you and the water push against each other. In each case there is a pair of forces, one *action* and the other *reaction*, that make up a single *interaction*. The reaction forces are what account for our motion in these cases. These forces depend on friction; a person or car on ice, for example, may not be able to exert the action force to produce the needed reaction force (even standing on frictionless ice would be difficult). Which force we call *action* and which we call *reaction* doesn't matter. The point is that neither exists without the other.

Action: tire pushes on road   Reaction: road pushes on tire

Action: rocket pushes on gas   Reaction: gas pushes on rocket

Action: man pulls on spring   Reaction: spring pulls on man

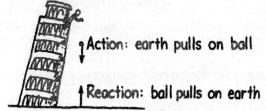

↑Action: earth pulls on ball

↑Reaction: ball pulls on earth

My daughter Leslie teaches in an elementary school and invokes the central concepts of physics with very young children. The lesson of the third law is comprehended by her students who learn that forces occur in pairs, that when one thing pushes on another, it itself is pushed. She makes this point with toy magnets. When one magnet moves another, it is also moved by the other. For equal mass magnets, the effect is most noticeable. That's because changes in motion are the same for each. For different size magnets, she gets them to see that smaller masses move more. The central concepts of physics can be understood by children. When they're older and take science classes, the central concepts are often obscured by the rigor that usually accompanies them.

Question

The strong man can withstand the tension force exerted by the two horses pulling him in opposite directions. How would the tension compare if only one horse pulled and the left rope were tied to a tree? How would the tension compare if the two horses pulled in the same direction, with the left rope tied to the tree?

Systems of forces*

An interesting question often arises: since action and reaction forces are equal and opposite, why don't they cancel to zero? To answer this question we must consider the *system* involved. Consider the force pair between the apple and orange in the sketch. Suppose we ignore the apple and everything else and consider only the orange. We can draw an imaginary circle around the orange and call what is within the circle the system. The pull from the apple supplies a force on the system and the system accelerates. Here the interaction is between the system (the orange) and something external (the

Answer

The tension is the same if the tree takes the place of one horse! The tree "holds" the rope just as the horse would. If two horses pull in the same direction—ouch! Rope tension doubles. (It doubles whether the tree or another pair of horses hold the left end!)

---

* If you wish to avoid a deeper plow setting, you may want to skim through this section. The idea, however, will occur again when we treat momentum conservation in the next chapter.

apple), so the action and reaction forces don't cancel. The fact that the orange simultaneously exerts a force on the apple, which is external to the system, may affect the apple but not the orange.

If, however, we consider the system to be both the orange and apple together, the force pair is internal to the system. In this case the forces do cancel each other. If there were no friction forces or other forces acting on the system, the apple and orange would move closer but the system's "center of mass" would be in the same place before and after the pulling. (We'll return to this idea of systems in the next chapter, and to center of mass in Chapter 6.) The system itself would not accelerate. Similarly, the many force pairs between molecules in a golf ball may hold the ball together into a cohesive solid, but they play no role at all in accelerating the ball. A force external to the ball is needed to accelerate the ball. Similarly, a force external to both the apple and orange is needed to produce acceleration of both (like friction of the floor on the apple's feet). If this is confusing, it may be well to note that Newton had difficulties with the third law himself.

Questions

1. On a cold rainy day your car battery is dead and you must push the car to get it started. Why can't you push the car by remaining comfortably inside and pushing against the dashboard?
2. Does a speeding missile possess force?
3. Which force is greater, the earth's pull on the moon, or the moon's pull on the earth?
4. Can you identify the action and reaction forces in the case of an object falling in a vacuum?
5. Suppose two carts, one twice as massive as the other, fly apart when the compressed spring that joins them is released. How fast does the heavier cart move compared with the lighter cart?

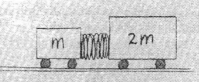

Acceleration and Newton's Third Law

When a rifle is fired there is an interaction between the rifle and the bullet. A pair of forces act on both the rifle and the bullet. The force exerted on the bullet is as great as the reaction force exerted on the rifle; hence, the rifle kicks. Since the forces are equal in magnitude, why doesn't the rifle recoil with the same speed of the bullet? In analyzing changes in motion, Newton's second law

reminds us that we must also consider the masses involved. Suppose we let F represent both the action and reaction force, m the mass of the bullet, and M the mass of the more massive rifle. The accelerations of the bullet and the rifle are then found by taking the ratio of force to mass. The acceleration of the bullet is given by

$$\frac{F}{m} = a$$

while the acceleration of the recoiling rifle is

$$\frac{F}{M} = a$$

We see why the change in motion of the bullet is so huge compared to the change of motion of the rifle. A given force divided by a small mass produces a large acceleration, while the same force divided by a large mass produces a small

Answers

1. In this case the system to be accelerated is the car. If you remain inside and push on the dash, the force pair you produce acts and reacts within the system. These forces cancel out as far as any motion of the car is concerned. To accelerate the car there must be an interaction between the car and something else—you pushing outside against the road, for example.
2. No, a force is not something an object has, like mass, but is part of an interaction between one object and another. A speeding missile may possess the capability of exerting a force on another object when interaction occurs, but it does not possess force as a thing in itself. A speeding missile possesses speed, and as we will see in the following chapter, it also possesses momentum.
3. As my friend Charlie Spiegel says, this is like asking which is greater: the distance between New York and Los Angeles or the distance between Los Angeles and New York. Just as there's one distance with two opposite directions between these cities, there's one interaction with two equal and opposite forces between the earth and the moon.
4. To identify a pair of action-reaction forces in any situation, first identify the pair of interacting objects involved. Something is interacting with something else. In this case the whole earth is interacting (gravitationally) with the falling object. So the earth pulls downward on the object (call it action), while the object pulls upward on the earth (reaction). The earth and object pull on each other.
5. In accord with Newton's third law, the force on each will be the same. But since the masses are different, accelerations will differ. The twice-as-massive cart will undergo half the acceleration of the less massive cart and will gain only half the speed.

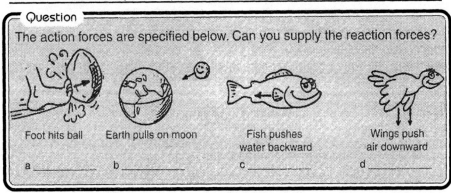

The action forces are specified below. Can you supply the reaction forces?

Foot hits ball　　Earth pulls on moon　　Fish pushes water backward　　Wings push air downward

a _____　b _____　c _____　d _____

acceleration. We use different-sized symbols to indicate the differences in relative masses and resulting accelerations.

Rocket Propulsion

If we extend the idea of a rifle recoiling or "kicking" from the bullet it fires, we can understand rocket propulsion. Consider a machine gun recoiling each time a bullet is fired. If the machine gun is fastened so it is free to slide on a vertical wire, it accelerates upward as bullets are fired downward. A rocket accelerates the same way as it continually "recoils" from the ejected exhaust gas. Each molecule of exhaust gas is like a tiny bullet shot from the rocket.

It is a common misconception that a rocket is propelled by the impact of exhaust gases against the atmosphere. In fact, at the beginning of the century, before the advent of rockets, it was commonly thought that sending a rocket to the moon was impossible because of the absence of an atmosphere for the rocket to push against. But this is like saying a gun wouldn't kick unless the bullet had air to push against. Not true! Both the rocket and recoiling gun accelerate not because of any pushes on the air, but because of the reaction forces by the "bullets" they fire—air or no air. A rocket works better, in fact, above the atmosphere where there is no air drag to restrict its speed.

Air Flight

Using Newton's third law, we can understand how a helicopter gets its lifting force. The whirling blades are shaped to force air particles down (action), and

Answer

(a) Ball hits foot; (b) Moon pulls on Earth; (c) Water pushes fish forward;
(d) Air pushes wings upward

the air forces the blades up (reaction). This upward reaction force is called *lift*. When lift equals the weight of the craft, the helicopter hovers in midair. When lift is greater, the helicopter climbs upward.

This is true for birds and airplanes. When a bird is soaring, its wings must be shaped so that moving air particles are deflected downward. Birds fly by pushing air downward. The air simultaneously pushes the bird upward. The air pushed downward by a bird often meets the air below and swirls upward, which creates an updraft strongest off to the side of the bird. Geese and ducks take advantage of this, and position themselves to get added lift from the updraft. Each bird in turn creates an updraft for the following bird, and so on. So we see why geese and ducks fly in a V formation.

Slightly tilted wings that deflect oncoming air downward produce the lift on an airplane.* Air must be pushed downward continuously to maintain lift. This supply of air is obtained by the forward motion of the aircraft, which results from jets that push air backward. When the jets push air backward, the air pushes the jets forward. Lift of an aircraft or of a bird is enhanced by the wing's curved surface and resulting airfoil, which is another story discussed in Book Two. So we see that explanations often involve more than a single concept.

We see Newton's third law at work everywhere. A fish pushes the water backward with its fins, and the water pushes the fish forward. The wind pushes against the branches of a tree, and the branches push back on the wind and we have whistling sounds. Forces are interactions between different things. Every contact requires at least a two-ness or pair; there is no way that an object can exert a force on nothing or that nothing can exert a force on something. Forces, whether large shoves or slight nudges, always occur in pairs, each of which is opposite to the other. Thus, we cannot touch without being touched.

Summary of Newton's Three Laws

An object at rest tends to remain at rest; an object in motion tends to remain in motion at constant speed along a straight-line path. This tendency of objects to resist change in motion is called *inertia*. Mass is a measure of inertia. Objects will undergo changes in motion only in the presence of a net force.

When a net force acts on an object, the object will accelerate. The acceleration is directly proportional to the net force and inversely proportional to

the mass. Symbolically, $a \sim \dfrac{F_{net}}{m}$. Acceleration is always in the direction of the

---

* Hold your hand like a flat wing outside the window of a moving automobile. Then slightly tilt the front edge upward so air is deflected downward. Notice the lifting effect. Can you see Newton's laws at work here?

net force. When objects fall in a vacuum, the net force is simply the weight, and the acceleration is g (the symbol g denotes that acceleration is due to gravity alone.) When objects fall in air, the net force is equal to the weight minus the force of air drag, and the acceleration is less than g. If and when the force of air drag equals the weight of a falling object, acceleration terminates, and the object falls at constant speed (called terminal speed).

Whenever one object exerts a force on a second object, the second object exerts an equal and opposite force on the first. Forces come in *pairs*, one action and the other reaction, both of which constitute the interaction between one thing and the other. Neither force exists without the other.

> **Question**
>
> A high-speed bus and an innocent bug have a head-on collision. The force of the bus on the bug splatters the bug over the windshield. Is the corresponding force that the bug exerts against the windshield greater, less, or the same? Is the resulting deceleration of the bus greater than, less than, or the same as that of the bug?

Newton's laws of motion are as valid today as they were before the time of Einstein and other 20th-century physicists. Newton's laws are all that is needed to compute the motion of projectiles near the earth's surface with an accuracy of better than 1 part in 100 million. Newton's laws were quite sufficient in putting men on the moon. Although refinements to Newton's laws occur in the domains of the very small (quantum mechanics), very fast (relativity), and very large (astrophysics), Newton's laws provide a framework for understanding the physics of these extremes.

> **Answer**
>
> The magnitudes of both forces are the same. They constitute an action-reaction force pair that makes up the interaction between the bus and the bug. The accelerations, however, are very different because the masses are different. The bug undergoes an enormous and lethal deceleration, while the bus undergoes a tiny deceleration—so tiny the passengers won't notice the slight slowing. If the bug were a more massive thing—another bus, for example—the slowing down would be quite evident!

4

Momentum

Studying physics concepts such as forces and motion today is much easier than the study of these concepts in the past. That's because we benefit from 20/20 hindsight and refinement of the original descriptions and explanations of concepts. Newer treatments of concepts are often (though not always!) simpler and more straightforward. Newton's laws are an example. In the previous chapter we stated that Newton's second law is expressed by the relationship, $a = F/m$. This expression, however, is a refinement of the way Newton presented it. Newton expressed it: $F = \Delta p/\Delta t$, which is read, "The force impressed on an object will be equal to the time rate of change in momentum of the object." Momentum is inertia in motion, like a huge truck loaded with lead storage batteries rolling without brakes down a long steep hill. Momentum is oomph. More specifically,

$$\text{Momentum} = \text{mass} \times \text{velocity}$$

In shorthand notation,

$$p = mv$$

where p is momentum, m is mass of the body being moved, and v is the velocity of the body. Bold letters are often used for p and v ($\mathbf{p} = m\mathbf{v}$) to indicate they are vector quantities. Later in this chapter we'll see that the directional property of momentum has fascinating implications. When direction is not an important factor, we can say momentum = mass × speed, which we still abbreviate mv. For now we'll focus on how momentum can be changed.

Impulse Changes Momentum

To change the momentum of something requires force and, very importantly, duration of the force—time. Apply a force briefly to a stalled automobile, and you produce a small change in its momentum. Apply the same force over an extended time interval, and a greater change in momentum results. We name the product of force and this time interval *impulse*. The relationship of impulse to momentum is a rearrangement of the Chapter Three version of Newton's second law ($a = F/m$). The time interval of impulse is buried in the term for acceleration (change in velocity/time interval). Rearrangement of Newton's second law gives*

Force × time interval = change in (mass × velocity)

We can express all terms in shorthand notation and introduce the delta symbol Δ (Greek letter D) which stands for "change in" (or "difference in"):

$$F\Delta t = \Delta mv$$

F means *average* force, for force usually varies over Δt. The equation reads, "Average force multiplied by the time-during-which-it-acts equals change in momentum."

This rearrangement of Newton's second law explains why follow-through is important in increasing the momentum of things. You apply the largest force you can, and apply it for as long a time as you can to impart maximum velocity to something. This shows why the long barrels of cannons increase the velocity of emerging cannonballs. The force of exploding gunpowder in a long barrel acts on the cannonball for a longer time. The longer the time that the force acts, the greater is the increase in momentum.

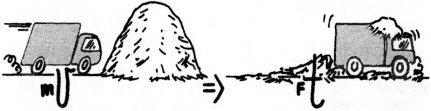

Correspondingly, if we decrease momentum over a long time, a smaller force results. A car out of control is better off hitting a haystack than a brick wall. By hitting the haystack the car may extend its time of impact 100 times. Then the force of impact is reduced by 100. So whenever you wish the force of impact to be small during a collision, extend the time of impact. The safety net used by acrobats is an obvious example of a long duration to reduce the momentum of a fall. Another example is catching a fast baseball with your bare hand. In doing so, you extend your hand forward so you'll have plenty of room to let your hand move backward after you make contact with the ball. You extend the time of impact and thereby reduce the force of impact. Similarly, a boxer rides or rolls with the punch to reduce the force of impact.

My sport was boxing in my teen years. I was tall for my

---

* If we equate the cause of acceleration ($F/m = a$) to what acceleration is ($a = \Delta v/\Delta t$), simple rearrangement gives $F\Delta t = m\Delta v$ or $F\Delta t = \Delta mv$.

weight (112 pounds and only two inches shy of six feet) so I had the advantage of reach over my opponents. I knew nothing of physics, but nevertheless had an explanation for

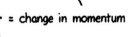
F t = change in momentum

F t = change in momentum

why one experiences a less forceful punch when riding with the punch. I thought it had to do with a decrease in the relative speed of impact. This explanation was challenged when I learned that some boxers could throw a punch at a speed of ninety miles per hour. One moves back at about two miles an hour when riding with a punch, which would reduce the ninety miles per hour to eighty- eight miles per hour. But that's not much different, so the explanation had to be something else. Of course now I know that *speed* isn't the deciding factor—*time* is. Another misconception I had in my days of pugilism was the explanation of why I could spend more than an hour punching a heavy bag in the gym without tiring, yet during a bout, the three three-minute rounds were very tiring. I had attributed this to excess mental energy expended in the excitement and tenseness of the encounter with an opponent. My view of this is very different now: the reason I could hit the bag

without tiring was because whatever momentum I put into my punches was countered by the impulse of the bag upon my fists. The bag supplied the stopping impulse. But punches that missed when boxing with an opponent were stopped not by my opponent, but by myself. I unknowingly supplied the impulse to stop punches that didn't connect with my opponent. A boxer who misses many punches is a boxer who finds himself tiring quickly.

A popular classroom activity is having students throw raw eggs into a sagging sheet. The stopping time is long enough so that the eggs don't break. The converse idea of short time of contact explains how a karate expert can sever a stack of bricks with a bare-hand blow. Swift execution makes the time of contact very brief and correspondingly makes the force of impact huge.

The impulse-momentum relationship is put to a thrilling test in bungee jumping. When your fall is brought to a halt by the cord, you'll be glad the cord stretches, for whatever momentum you gain in your fall, the cord will have to supply the same impulse to bring you to a halt—hopefully above ground level.

Note how $F\Delta t = \Delta mv$ applies here. The mv you wish to change is the momentum you have gained before the cord does its thing. The $F\Delta t$ is the impulse the cord will supply to reduce your mv to zero. Because the cord takes so long stretching, a large Δt insures a correspondingly small force F on you. (Would you like to try such a jump with a non-stretchable cord attached to your feet? Good-bye feet!) Bungee cords typically stretch to about twice their length, depending on momentum.

During my prospecting days in Colorado, I lived near a railroad yard. Long trains of freight cars were switched from one track to another in the yard. It was bothersome when trying to sleep at night to hear the cars clanking against one another when the train was brought up to speed, or slowed down. A train of forty or more cars would make forty or more clanks when first being moved. Think about the great force required to change the momentum of a long line of loaded cars. A locomotive would be hard pressed to bring all forty or more cars up to speed simultaneously. Its wheels would much prefer to slip and squeal rather than grab the track. But the locomotive can change the speed on one car easily. And once that car is moving, it can with ease get another car moving. And so on. Momentum is changed one car at a time, by successive durations that yield increased impulse for a given force. Similarly for braking and slowing down. I tie this idea to the academic

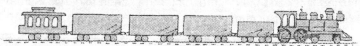

load students take in school. They can't handle all they're expected to learn in one semester, so they spread their courses out, a few at a time rather than all at once. They appreciate seeing ideas of physics applied to more than their physical surroundings!

So we see that the impulse-momentum relationship helps us to analyze many examples in which forces act and motion changes. Sometimes the impulse can be considered to be the cause of a change of momentum. Sometimes a change of momentum can be considered to be the cause of an impulse. It doesn't matter which way we think about it. The important thing is that impulse and change of momentum are always linked.

> **Question**
>
> When does impulse equal momentum?

Bouncing

You know that if a flower pot falls from a shelf onto your head, you may be in trouble. If it bounces from your head, you're much more likely to be in trouble. Impulses are greater when bouncing takes place. This is because the impulse required to bring something to a stop and then, in effect, "throw it back again" is greater than the impulse required merely to stop it. My pen-pal, Marilyn Hromatko, cites her experience playing volleyball to support this idea. She was instructed to stop an oncoming ball with her fingertips and then let another player hit the ball. Why? Because the impulse applied by two players was greater than she could muster to stop the ball and slam it back. So if we catch the falling pot with our hands, we provide an impulse to catch it and reduce its momentum to zero. Then if we throw the pot upward, we'd have to provide additional impulse. So we see that it would take more impulse to catch it and throw it back up than merely to catch it. The same greater impulse is supplied by your head if the pot bounces from it.

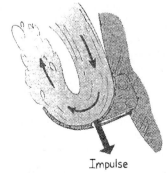

Impulse

An interesting application of the greater impulse that occurs when bouncing takes place was employed with great success in California during the gold rush

> **Answer**
>
> Generally, impulse equals a *change* in momentum. If the initial momentum of an object is zero when the impulse is applied, then impulse = momentum.

days. The water wheels used in gold-mining operations were inefficient. Lester A. Pelton saw that the problem had to do with their flat paddles, so he designed curve-shaped paddles that would cause the incident water to make a U-turn upon impact—to "bounce." In this way the impulse exerted on the water wheels was greatly increased. Pelton patented his idea and earned much more money from his invention, the Pelton wheel, than most of the miners made from gold.

Questions

1. When you deliver a karate chop to a stack of bricks, how will the impulse differ if your hand bounces back upon striking the bricks?
2. How does the force exerted on the bricks compare to the force exerted on your hand?

Conservation of Momentum

Newton's second law tells us that if we want to accelerate an object, we must apply a force. The force must be an external force. If you want to accelerate a car, you must push from outside. Inside forces don't count—sitting inside an automobile pushing against the dashboard with the dashboard pushing back has no effect in accelerating the automobile. Likewise, if we want to change the momentum of an object, the force must be external. In the impulse-momentum concept, $F\Delta t = \Delta mv$, internal forces don't count. If no external force is present, then no change in momentum is possible.

Consider a rifle be-
ing fired. Both the force
that drives the bullet and
the force that makes the
rifle recoil are equal and
opposite (Newton's third
law). To the system com-

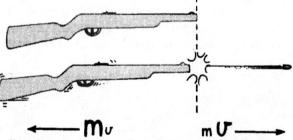

prised of the rifle and the bullet, they are internal forces. No external net force acts on the rifle-bullet system so the total momentum of the system undergoes no net change. Before the firing, the momentum is zero; after firing, the *net* momentum is still zero. Like velocity, momentum is a *vector quantity*. The momentum gained by the bullet is equal and opposite to the momentum

---

Answers

1. The impulse will be greater if your hand bounces upon impact. If the time of impact is not correspondingly increased, a greater force will be exerted on the bricks (and your hand!).
2. In accord with Newton's third law, the forces will be equal. Only the resilience of your hand, and its toughening by training, allows this feat without broken bones.

gained by the rifle. They cancel. No momentum in the rifle-bullet system is gained and none is lost.

In the rifle-bullet example, we have neglected the momentum of the gases produced by the exploding gunpowder. More than one person has unfortunately overlooked the momentum of ejected gunpowder gases when firing blanks. Being shot by a blank when the gun is several feet away is okay—but not when the gun is close. Actor John Eric Hexum in 1992 accidentally killed himself when he held a pistol loaded with a blank bullet to his head and fired it. Bruce Lee's son was similarly killed by a shot in the stomach from a closely held gun firing a blank. Although no slug emerges from the gun when fired, exhaust gases do. So strictly speaking, the system being considered is comprised of the rifle, bullet, and gunpowder. We should say that the momentum of the bullet + gunpowder gases is equal and opposite to the momentum of the recoiling rifle.

Questions

1. If you toss a ball horizontally while on a skateboard, you'll roll backward with the same amount of momentum given to the ball. Will you roll backward if you go through the motions of tossing the ball, but instead hold onto it?

2. In the previous chapter we questioned the relative speeds of a pair of carts, one twice the mass of the other, after a connecting compressed spring was released. In terms of momentum conservation, how do the speeds of the carts compare?

We spoke briefly in the previous chapter about what we mean by *system*. Consider a cue ball that makes a head-on collision with an 8-ball at rest. We know what happens: the cue ball stops dead in its tracks when it hits the 8-ball, and the 8-ball continues with the speed that the cue ball had before collision. Every pool player knows this. Now let's analyze the situation and concern ourselves only with the 8-ball. We can draw an imaginary line around the 8-ball and let only it be our system—Case I on the next page. Our system is initially at rest. Then the cue ball exerts a force on our system and it suddenly moves. An impulse on the system, the 8-ball, changes its momentum—$F\Delta t = \Delta mv$. Pretty straightforward.

Answers

1. No, you'll not roll back without immediately rolling forward to produce no net rolling. In terms of momentum, if no net momentum is imparted to the ball, no net momentum will be imparted to you. Or in third-law fashion, if no net force acts on the ball, no net force acts on you.

2. By $(2m \times v) = (m \times ?)$, we see that the half-as-massive cart will have twice the speed of the more massive cart.

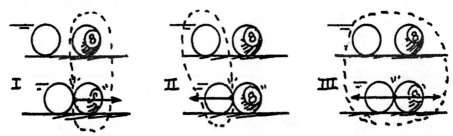

Now let's concern ourselves only with the cue ball. We can draw an imaginary line around it, and consider the cue ball to be our system—Case II above. Initially it's moving; it has momentum. Then when it makes impact with the 8-ball, it experiences a force (the reaction to the force it exerts on the 8-ball). This is a force from outside the system, which over the time of contact comprises an *impulse*. What does the impulse do? It changes the momentum of the cue ball and brings it to rest. The impulse on the cue ball has the same magnitude as the impulse on the 8-ball. Both systems underwent the same change in momentum.

Now for a more interesting way to look at the collision, consider the single system comprised of both balls. Draw an imaginary line around both balls—before collision, during collision, and after collision—Case III above. Before collision, the system has momentum. It's all in the cue ball. When the balls collide, is there an external force? The answer is no. The force pair that acts on the balls during collision is internal to the system. When the F is zero in $F\Delta t = \Delta mv$, then impulse is zero and the change in momentum is zero. We're not saying the momentum is zero, but the CHANGE in momentum is zero. Put another way, in the absence of an external force, no change in momentum occurs. In fact, no change in momentum is possible. This brings us to one of the cornerstone principles of physics—the conservation of momentum.

In the absence of an external force, the momentum of a system cannot change.

I use a practice sheet to get across the idea of a system with my students. A modified version is shown on the facing page. Try this Practice Page yourself. The answers are displayed at the end of this chapter on page 77.

Whenever a physical quantity remains unchanged during a process, that quantity is said to be *conserved*. We say momentum is conserved.

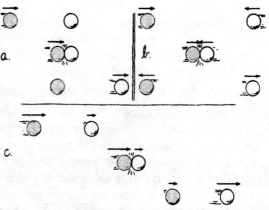

The law of momentum conservation is especially useful in collisions, where the forces involved are internal forces. In any collision, we can say

Net momentum before collision = net momentum after collision.

This is true whether the collision is elastic or inelastic. When objects collide without lasting deformation or without the generation of heat, the collision is said to be an **elastic collision**. Perfectly elastic collisions occur when molecules in a gas collide with one another. Colliding billiard balls only approximate a perfectly elastic collision. Collisions of macroscopic objects are partly elastic and partly inelastic.

An **inelastic collision** is characterized by deformation or the generation of heat, or both. Momentum is conserved whether a collision is elastic or inelastic. A

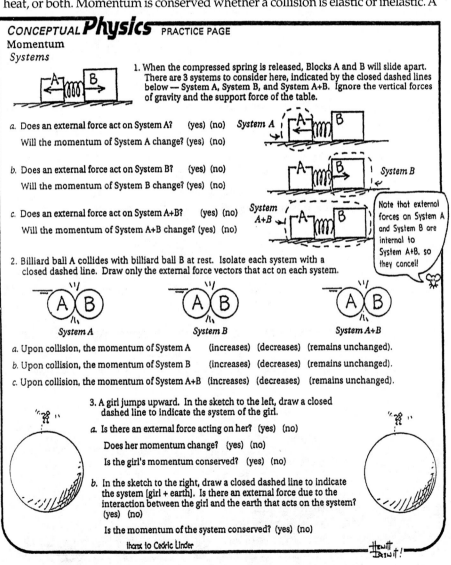

CONCEPTUAL *Physics* PRACTICE PAGE

Momentum
Systems

1. When the compressed spring is released, Blocks A and B will slide apart. There are 3 systems to consider here, indicated by the closed dashed lines below — System A, System B, and System A+B. Ignore the vertical forces of gravity and the support force of the table.

a. Does an external force act on System A?   (yes) (no)

  Will the momentum of System A change? (yes) (no)

b. Does an external force act on System B?   (yes) (no)

  Will the momentum of System B change? (yes) (no)

c. Does an external force act on System A+B?   (yes) (no)

  Will the momentum of System A+B change? (yes) (no)

Note that external forces on System A and System B are internal to System A+B, so they cancel!

2. Billiard ball A collides with billiard ball B at rest. Isolate each system with a closed dashed line. Draw only the external force vectors that act on each system.

System A   System B   System A+B

a. Upon collision, the momentum of System A   (increases) (decreases) (remains unchanged).

b. Upon collision, the momentum of System B   (increases) (decreases) (remains unchanged).

c. Upon collision, the momentum of System A+B (increases) (decreases) (remains unchanged).

3. A girl jumps upward. In the sketch to the left, draw a closed dashed line to indicate the system of the girl.

a. Is there an external force acting on her? (yes) (no)

  Does her momentum change? (yes) (no)

  Is the girl's momentum conserved? (yes) (no)

b. In the sketch to the right, draw a closed dashed line to indicate the system [girl + earth]. Is there an external force due to the interaction between the girl and the earth that acts on the system? (yes) (no)

  Is the momentum of the system conserved? (yes) (no)

thanx to Cedric Linder

Questions

1. Consider the air track shown to the right. Suppose a gliding cart bumps into and sticks to a stationary cart that has three times the mass. Compared to the initial speed of the gliding cart, how fast will the coupled carts glide after collision?

2. A hungry 4-kg fish swims toward and encounters a sleepy 1-kg fish swimming toward it at four times its own speed. Compared to before lunch, how fast is the hungry fish moving immediately after lunch?

collision is completely inelastic when the colliding objects become entangled during the collision. The energy (next chapter) of such a collision is transformed to energy of another kind—usually thermal energy. Consider, for example, the case of a freight car moving along a track and colliding with another freight car at rest. If the freight cars are of equal mass and are coupled by the

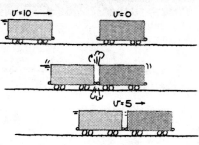

collision, can we predict the velocity of the coupled cars after impact?

Suppose the single car is moving at 10 m/s, and we consider the mass of each car to be m. Then, from the conservation of momentum,

$$(\text{Total } mv)_{\text{before}} = (\text{total } mv)_{\text{after}}$$
$$(m \times 10)_{\text{before}} = (2m \times V)_{\text{after}}$$

By simple algebra, $V = 5$ m/s. This makes sense, for since twice as much mass is moving after the collision, the velocity must be half as much as the velocity before collision. Both sides of the equation are then equal.

Answers

1. According to momentum conservation, momentum of cart of mass m and velocity v before = momentum of both carts stuck together after.

$$mv_{\text{before}} = (m + 3m)\, V_{\text{after}}$$
$$mv = 4mV$$

$$V = \frac{mv}{4m} = \frac{v}{4}$$

This makes sense, for four times as much mass will be moving after collision, so the coupled carts will glide more slowly. No momentum loss means four times the mass glides 1/4 as fast.

2. Each fish has equal and opposite momenta, so the two-fish system has zero momentum before and after lunch. The hungry fish is brought to a halt.

The conservation of momentum is nicely illustrated on a pool or billiards table—not only in head-on collisions, but on glancing shots as well. When angles are involved, the vector nature of momentum is evident. No matter how complicated the collision of balls, the momentum along the line of action of the cue ball before impact is the same as the combined momentum of all the balls along this same line after impact. Also, the components of momenta perpendicular to this line of action cancel to zero after impact, the same value as before impact in this direction. This is more clearly seen when rotational skidding, English, is not imparted to the cue ball. When English is imparted by striking the cue ball off center, rotational momentum—which is also conserved—somewhat complicates analysis. But regardless of how the cue ball is struck, in the absence of external forces, both linear and rotational momentum are always conserved. Pool or billiards offers a first-rate exhibition of momentum conservation in action. In my textbooks, I suggest as a project that students cut classes when they're a bit ahead in their studies and visit a billiards or pool parlor to put their knowledge of momentum to a test. The qualifier, "when a bit ahead in their studies" ensures that hordes of students aren't urged to head for the pool parlors. In my college-student days, I was never ahead in my studies. I remember always being behind, which detracted greatly from thoroughly enjoying school. Christmas and spring breaks were "catch-up" times rather than vacations.

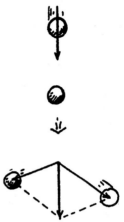

I first heard the eminent physicist Richard Feynman speak at an American Association of Physics meeting in San Francisco in 1972. Feynman explained how physicists investigate the innards of atoms by comparing an atom to a watch. If you wish to know what's inside a watch, one way to find out is to throw it as hard as you can against a cement wall. When it shatters, take photographs of the pieces as they scatter. With the rules of momentum conservation you can then deduce the relative masses of the inner parts. Likewise with atoms, where the tracks of subatomic particles are revealed in a variety of tracking chambers. Momentum conservation enables researchers to learn about the masses and electric charges of subatomic particles without knowing any details about the interaction forces that act in the collision.

Question

When a stationary uranium nucleus undergoes fission, it breaks into two unequal chunks that fly apart. What can you conclude about the momenta of the chunks? What can you conclude about the relative speeds of the chunks?

Answer

The chunks have equal and opposite momenta, since their total momentum must remain zero. The more massive chunk has less speed.

Richard Feynman was always a hero of mine. I met him in November, 1987, when we were on a panel at an AAPT meeting in Pasadena in Southern California. My friend Charlie Spiegel had given him the first edition of my high school version of *Conceptual Physics* and I was honored that Feynman read it, flattered that he liked it, and elated by the good advice he suggested for the second edition. The panel was comprised of six people, including Feynman and myself, and the topic was "what physics should be taught in high school." I had just finished my high school book and I stated my opinion that physics should be taught to all students, preferably in the 9th or 10th grade, to be the first course in a science sequence. Feynman, on the other hand, stated that physics has been so poorly taught in high school for so many years that increasing enrollments would make things worse. He felt too many misconceptions were taught, and no physics was better than poorly-taught physics. So here we were, on opposite sides of the fence. How ironic to be on a panel with my hero and find myself in disagreement with him! I was convinced that if Feynman had seen my whole program and not just the textbook, that he would agree that teaching the program would address many of his concerns about misconceptions. He and others stated that there weren't enough teachers to service all students. I countered that the many teachers needed were presently in our classrooms, or soon to be, and that a conceptual orientation that didn't overkill the math problem segment of physics would save the day. I didn't have the chance to confer more with Feynman. He died 3 months later. That meeting was his last public appearance. As I was grieving Feynman's passing, for I loved the guy, my son James was killed in a car accident. 1988 was a tough year for me.

But life goes on. Conservation of momentum and, as we shall see in the next chapter, conservation of energy, are the two most powerful tools of mechanics. Applying them yields detailed information that ranges from facts about the interactions of subatomic particles to the structure and motion of entire galaxies.

Questions

1. You can survive a feet-first impact at a speed of about 12 m/s (27 mi/h) on concrete; 15 m/s (34 mi/h) on soil; and 34 m/s (76 mi/h) on water. Why the different values for different surfaces?

2. Which would be more damaging: driving into a massive concrete wall or having a head-on collision with an identical car traveling toward you at the same speed?

Answers

1. The duration in which your change of momentum occurs is different on different surfaces.

2. Both cases are equivalent. Either way, your car rapidly decelerates to a dead stop. If the oncoming car were traveling slower with less momentum, you'd keep going after collision with more "give" and less damage to you. And if the oncoming car had more momentum than you, it would keep going and you'd snap into a sudden reverse with greater damage. Identical cars at equal speeds means equal momenta—zero before, zero after collision.

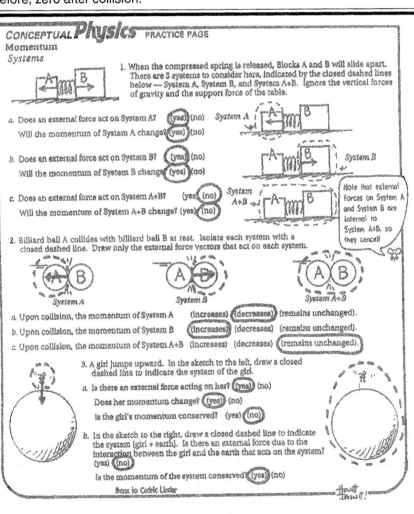

5

Energy

When I teach about energy at City College of San Francisco, I employ a long length of wire that hangs from a high ceiling above the lecture table. I fashion it into a pendulum with a massive bowling ball at its end. I say nothing as I pull the ball to the side, stand on a chair with my head against the wall, hold the ball in front of my teeth, then release it. You can hear a pin drop as the class watches the ball swing across the room, come to a momentary stop, then swing back toward me. With eyes closed, I don't flinch as it comes very close to its starting point. I break my silence in saying, "I believe in the conservation of energy!" This demo is good not only for its potential violence, but more important, for the visual evidence it presents for potential energy changing to kinetic energy and back again.

Energy is perhaps the most central concept to all of science. It comes to us in the form of electromagnetic waves from the sun and we feel it as heat; it is captured by plants and binds atoms of matter together; it is in the food we eat and we receive it through metabolism. But what *is* energy?

All through my college-student days and into my teaching years I have been bothered about defining energy. Unlike concepts like acceleration and momentum, which are clearly defined and seem to make good sense, the concept of energy is more elusive. I could never put my finger on exactly what energy is. So when I started writing *Conceptual Physics* after teaching for five years, I decided I'd delve into and settle the energy question in my own mind, then spell it out clearly and definitively for my students and readers. If you want to be clear about an idea or concept, put it in writing. Holes soon become apparent. And so it was with energy—I couldn't spell it out like other concepts.

Before you can explain an idea or concept to somebody, you've got to be sure you understand it yourself. Then, in explaining it, you've got to begin by building on knowledge already in the mind of the person you're teaching. No physics can be taught unless it relies on what is learned or on something the learner already knows.

My best shot at explaining energy now is to say that it is nature's way of keeping score. We sense energy only when the score changes, when there is a transformation from one form to another, or a transfer from one point to another—like when electromagnetic waves from the sun vibrate the cells in our skin, or when water from a high lake rushes through a turbine at a lower location. Energy is only apparent when it undergoes a change—either in form or location. Just as water that smoothly flows through a turbine is composed of tiny lumps—H_2O molecules, energy changes also occur in tiny lumps—*quanta*. We'll study quanta in Book Four when we study light. For the larger-scale energy changes treated in this chapter, we'll ignore the lumpy nature of energy. We'll begin our study of energy by learning about a relatively simple and related concept: work.

Work

The word work has several meanings in our everyday world, such as the opposite of play. In science, however, the word *work* has a specific meaning. We say **work** is defined as the quantity "force multiplied by distance." For example, when we lift a load against earth's gravity, we say work is done on the load. The heavier the load or the higher we lift the load, the more work is done. When we compress a spring, we do work on the spring. The greater the force required for compression, and the farther we compress the spring, the more work is done. Two things determine whether work is being done: (1) the exertion of a force and (2) the movement of something by that force. We define the work done on an object as the product of the applied force and the distance through which the object is moved by that force:*

$$\text{Work} = \text{force} \times \text{distance}$$

If we lift two loads one story upward, we do twice as much work as in lifting one load because the *force* needed to lift twice the weight is twice as much. Similarly, if we lift a load up two stories instead of one story, we do twice as much work because the *distance* is twice as much. How much work do we do if we lift twice the load twice as high? Do you see the answer is four times as much? If not, apply the above equation. If so, onward!

A weightlifter who holds a stationary heavy barbell overhead does no work *on the barbell*. The weight

---

*   Sometimes a force doesn't act in the same direction of movement. But some "fraction" of the force may; that is, a *component* of the force may act in the direction of motion. So more specifically, work is the product of the component of force that acts in the direction of motion and the distance moved.

lifter may get really tired holding the barbell up, but if the barbell is not moved by the force he exerts, he does no work on the barbell. Work may be done on the muscles by stretching and contracting, which is force times distance on a biological scale, but this work is not done on the barbell. *Lifting* the barbell, however, is a different story. When the weightlifter raises the barbell from the floor, he does work on the barbell.

We can express work numerically. If we know the force in newtons and the distance moved in meters, multiply both. This gives a number—the amount of work. The unit of work is the newton-meter (N· m), also called the *joule* (J) (rhymes with *cool*). One joule of work is done when a force of 1 newton is exerted over a distance of 1 meter, as in lifting an apple over your head. For larger values we speak of kilojoules (kJ), thousands of joules, or megajoules (MJ), millions of joules. A weight lifter typically does work in kilojoules. The energy released by one kilogram of gasoline is rated in megajoules.

Work Changes Energy

Watch a pile driver in action, and you're seeing a good example of work and its relationship to energy. Work is done in lifting the heavy ram of a pile driver. As a result, the ram acquires the property of being able to do work on a piling beneath it when it falls. We say the work done on the ram gives it energy. Similarly, when work is done by an archer in drawing a bow, the bent bow has the ability of being able to do work on the arrow. When work is done to wind a spring mechanism, the spring acquires the ability to do work on various gears to run a clock, ring a bell, or sound an alarm. In each case, something has been acquired. This "something" that is given to the object enables the object to do work. This "something" may be a compression of atoms in the material of an object; it may be a physical separation of attracting bodies;

it may be a rearrangement of electric charges in the molecules of a substance. This "something" that enables an object to do work is *energy*.* Like work, energy is measured in joules. Energy appears in many forms, which are featured in all branches of physics and the other sciences. For now we focus on mechanical energy—the form of energy due to position (potential energy) or to the movement of mass (kinetic energy). Mechanical energy may be in the form of either potential energy or kinetic energy. Let's consider them in turn.

---

*Strictly speaking, that which enables an object to do work is its *available* energy, for not all the energy in an object can be transformed to work—some unavoidably dissipates as heat.

Potential Energy

We've said that an object may store energy because of its position. The energy that is stored and held in readiness is called **potential energy** (PE), because in the stored state it has the potential to do work. For example, a stretched or compressed spring has potential energy because of its position, for if it is part of a popgun, it is capable of doing work.

The chemical energy in fuels is potential energy, for it is energy of position from a microscopic point of view. This energy is available when the positions of electric charges within and between molecules are altered, that is, when a chemical change takes place. Any substance that can do work through chemical action possesses potential energy. Potential energy is found in fossil fuels, electric batteries, and the food we eat.

The easiest-to-visualize form of potential energy occurs when work is done to elevate objects against earth's gravity. The potential energy due to elevated positions is called *gravitational potential energy*. Water in an elevated reservoir and the elevated ram of a pile driver have gravitational potential energy. The amount of gravitational potential energy possessed by an elevated object is equal to the work done against gravity in lifting it. The work done equals the force required to move it upward times the vertical distance it is moved ($W = Fd$). The upward force equals the weight of the object, so the work done in lifting it is weight × height.

Gravitational potential energy = weight × height

Interestingly, the height refers to the vertical distance above some reference level, such as the ground or the floor of a building. The potential energy is relative to that level and depends only on weight and the vertical height. The potential energy of the ball at the top of the three structures in the sketch is the same for each, because each is raised the same vertical height. Gravitational potential energy depends on vertical displacement and not on the path taken to get it there.

We can see this by comparing the amount of work needed to push a block up an inclined ramp and the amount of work needed to simply lift it the same

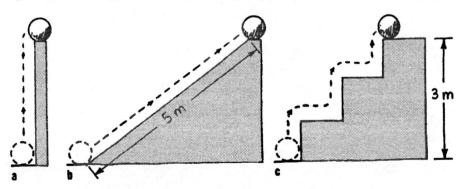

a   b   c   3 m

height. The force
needed to lift the
block straight up is its
weight. So the work
lifting is simply its
weight × height lifted.
The force needed to
push the block up an
incline is less than the

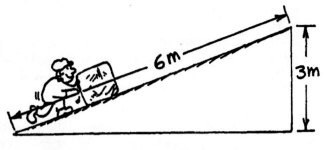

block's weight (which is why inclines are used). But the distance along the incline is correspondingly more. Less force × greater distance turns out to require the same work as simply lifting—if friction is negligible. As a numerical example, consider a block of very slippery ice that weighs 100 newtons pushed up a 6-meter long ramp inclined at 30°. At this angle, the height of the elevated end of the ramp is three meters (the hypotenuse of a 30°- 60°- 90° right triangle is twice the shorter side). By trigonometry, the force needed to slide the block up the 30° incline is half the weight of the block: 50 N. Work up the ramp is 50 N × 6 m = 300 J. Lifting the block requires 100 N × 3 m = 300 J. Either way, the work done increases the block's potential energy by 300 J. The ramp simply makes this work easier to perform.

Another example that illustrates this point is the displacement of a simple pendulum. Calculating the work done in displacing a simple pendulum bob is a typical problem for engineering physics students in their freshman year of college. Unlike the constant force needed to push a block up a straight-line incline, the force to move a pendulum bob along an arc varies with the distance moved. Calculating work when force varies with distance requires a form of mathematics called *integral calculus*. For example, consider a bowling ball suspended vertically by a length of rope. At the ball's lowest position, only a very slight sideways pull is needed to move it. As we continue pulling along the arc, the force increases. The punch line, to make a long story short, is that when the calculus is done, the work calculated is the same as the work to simply lift the ball the same vertical height. Again, potential energy depends on vertical displacement, and not on the path taken to get there. This simplicity of nature is quite amazing.

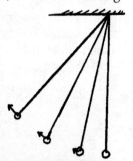

Questions

1. How much work is done on a bowling ball that weighs 80 newtons when you carry it horizontally across a 10-m-wide room?
2. How much work is done on it when you lift it 1 m?
3. What is its gravitational potential energy in the lifted position?

Answers

1. You do no work on the ball moved horizontally, since you apply no force except for the tiny bit to start it in its direction of motion. Evidence that no work was done on the ball is that it has no more PE across the room than it had initially. Work changes the energy of that which is worked on.
2. You do 80 J of work when you lift it 1 m (Fd = 80 N • m = 80 J).
3. It depends. Relative to its starting position, its potential energy is 80 J; relative to some other reference level, it would be some other value. If it were held over a deep well, for example, it would have a greater potential energy relative to the bottom of the well. The value of gravitational potential energy is always relative to some reference level.

Kinetic Energy

If we push on an object, we can set it in motion. More specifically, if we do work on an object, we can change the energy of motion of that object. If an object is moving, then because of that motion, it is capable of doing work. We call energy of motion **kinetic energy** (KE). The kinetic energy of an object depends on the amount of matter that moves and how fast it moves. Kinetic energy is equal to half the mass multiplied by the square of the speed.

$$\text{Kinetic energy} = \frac{1}{2}\text{mass} \times \text{speed}^2$$

A car moving along a road has kinetic energy. A twice-as-heavy car moving at the same speed has twice the kinetic energy. That's because a twice-as-heavy car has twice the mass. Kinetic energy depends on mass. But note how it also depends on speed—not just plain speed, but speed multiplied by itself—*speed squared*. If you double the speed of a car, you'll increase its kinetic energy by *four* (2^2 = 4). Or drive three times as fast and you have *nine* times the kinetic energy (3^2 = 9). The fact that kinetic energy depends on the square of the speed means that small changes in speed can produce large changes in kinetic energy. Got it? Now let's relate this to work.

We acquired the kinetic energy in the first place by doing work. We had to exert a force on the car to get it moving. That force multiplied by distance moved equals the gain in kinetic energy of the car.

$$\text{Net force} \times \text{distance} = \text{kinetic energy}$$

$$Fd = \frac{1}{2}mv^2$$

The kinetic energy of a moving object is equal to the work done in bringing it from rest to that speed or to the work a moving object can do in being brought to rest.* Work done is equal to the change in kinetic energy. This is an important relationship in physics, and is called the **work-energy theorem**. We abbreviate "change in" with the delta symbol, Δ, and say

$$\text{Work} = \Delta \text{KE}$$

Work equals change in kinetic energy. The work in this equation is the *net* work—that is, the work based on the net force. If, for instance, you are pushing on an object and friction is also acting on the object, the change of kinetic energy is equal to the work done by the net force, which is your push minus friction. (In this case, only part of the *total* work that you do is going into the object's kinetic energy. The rest transforms into heat, which we'll discuss later.)

The work-energy theorem is a central concept of mechanics. It emphasizes the role of change. If there is no change in energy, then we know no work was done. Recall our previous example of the weight lifter and the barbell. Work was done *on the barbell* when its potential energy was being changed. But when it was held stationary, no further work was being done *on the barbell* as evidenced by no further change in its energy. Similarly, push against a crate on a factory floor, and if it slides, then you're doing work on it. If it doesn't, then you are not doing work on the crate. Put rollers beneath the crate and push again. Now your push does work on the crate. The amount of work done on the crate is matched by its gain in kinetic energy.

The work-energy theorem applies as well to decreasing speed. The more kinetic energy something has, the more work is required to stop it. Twice as much kinetic energy means twice as much work. We do work when we apply the brakes to slow a car. This work is the friction force supplied by the brakes multiplied by the distance the friction force acts.**

Interestingly, friction supplied by the brakes is the same whether the car moves slowly or quickly. Friction doesn't depend on speed. The variable is the *distance* of braking. Note what this means: a car moving twice as fast as another takes four times as much work to stop, and will therefore take four times as much distance to stop. Accident investigators are well aware that an automobile going 100 kilometers per hour has four times the kinetic energy it would

---

*This can be derived as follows. If we multiply both sides of $F = ma$ (Newton's Second law) by d, we get $Fd = mad$. Recall from Chapter Two that $d = 1/2\ at^2$. Substituting, we can say $Fd = ma(1/2\ at^2) = 1/2\ m(at)^2$; and substituting $v = at$, we get $Fd = 1/2\ mv^2$. Equations are systems of *connections* that follow the tools of logic. Equations tell us what we can and can't ignore. We can appreciate the value of equations whether we do or don't carry out detailed mathematical operations.

**Never mind that we could call this "negative work" because the force is opposite to the direction of motion—we don't care about that kind of stuff here. We simply want to establish that work, positive or negative, equals the change in kinetic energy. We'll leave complications and peripheral distinctions to a textbook. Also, we needn't distinguish between friction of brake pads against brake drums and friction of the tires on the road. It turns out that braking friction against the inner drum of the wheel × distance of drum contact will produce the same value as the force of friction of the tires on the road × distance car decelerates on the road: either way, work = ΔKE.

have at 50 kilometers per hour. This means a car going 100 kilometers per hour will skid four times as far when its brakes are applied as it would going 50 kilometers per hour. This is because speed is squared for kinetic energy.

Kinetic energy underlies other seemingly different forms of energy such as heat, sound, and light. Random molecular motion is sensed as heat: when

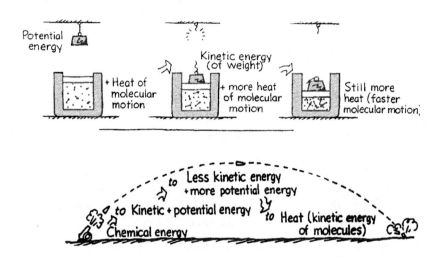

fast-moving molecules bump into the molecules in the surface of your skin, they transfer kinetic energy to your molecules much as balls in a game of pool or billiards. Sound consists of molecules vibrating in rhythmic patterns: shake a group of molecules in one place and in cascade fashion they disturb neighboring molecules that in turn disturb others, preserving the rhythm of shaking throughout the medium. Electrons in motion make electric currents. Even light energy originates from the motion of electrons within atoms. Kinetic energy is far reaching.

Conservation of Energy

More important than being able to state *what energy is* is understanding how it behaves—*how it is transformed*. Energy is evident only when it changes. For example, as we draw back the arrow in a bow, we do work in bending the bow: we give the arrow and bow potential energy. When released, most of this potential energy is transformed to the arrow as kinetic energy (the rest slightly warms the bow). The arrow in turn transfers this energy to its target, perhaps a bale of hay. The distance the arrow penetrates into the hay multiplied by the average force of impact doesn't quite match the kinetic energy of the arrow. The energy score doesn't balance. But if we investigate further, with sensitive thermometers, we'll find that both the arrow and hay are a bit warmer. By how much? By the energy difference! Energy changes from one form to another. Taking heat energy into account, we find energy transforms without net loss or net gain. Quite remarkable!

The study of various forms of energy and their transformations led to one of the greatest generalizations in physics—the law of **conservation of energy**:

> **Energy cannot be created or destroyed; it may be transformed from one form into another, or transferred from one place to another, but the total amount of energy never changes.**

When we consider any system in its entirety, whether it be as simple as a swinging pendulum or as complex as an exploding galaxy, without work input or work output there is one quantity that doesn't change: energy. It may change form or it may simply be transferred from one place to another, but, as far as we can tell, the total energy score stays the same. This energy score takes into account the fact that the atoms that make up matter are themselves concentrated bundles of energy. Mass itself is "energy of being," as stated in Einstein's celebrated equation, $E = mc^2$. Mass energy converts to radiant energy when the nuclei (cores) of atoms are rearranged. This occurs, for example, in the deep hot interior of the sun as gravitational pressure effectively jams the cores of hydrogen atoms together to form helium atoms. This welding together of atomic cores is called *thermonuclear fusion*. This process releases radiant energy, some of which reaches the earth. Part of this energy falls on plants, and part of this in turn later becomes coal or oil. Part of the energy from the sun goes into the evaporation of water from the ocean, and part of

Questions

1. Can you connect the concept of energy conservation to the fact that a ball has a higher speed when it rebounds from the "sweet spot" of a baseball bat or tennis racquet?

2. Does an automobile consume more fuel when its air conditioner is turned on? When its lights are on? When its radio is on while it is sitting in the parking lot?
3. Rows of wind-powered generators are used in various windy locations to generate electric power. Does the power generated affect the speed of the wind? Would locations behind the "wind mills" be windier if they weren't there?

this returns to the earth as rain that may become trapped behind a dam. By virtue of its position, the water in a dam has energy that may be used to power a generating plant below where it will be transformed to electric energy. This energy travels through wires to homes where it is used for lighting, heating, cooking, and operating electric gadgets. How nice that energy is transformed from one form to another!

Power

Energy can be transformed quickly or it can be transformed slowly. The rate at which energy is changed, or the rate at which work is done, is called **power**. We say power is equal to the amount of work done per time it takes to do it:

$$\text{Power} = \frac{\text{work done}}{\text{time interval}}$$

An engine of great power can do work rapidly. An automobile engine with

Answers

1. The sweet spot is the place where minimum vibration of the bar or racquet occurs when contact is made with the ball. Little or no vibration means little or no energy "wasted" in vibration, and more energy available for the ball.
2. The answer to all three questions is yes, for the energy they consume ultimately comes from the fuel. Even the energy from the battery must be given back to the battery by the alternator, which is turned by the engine, which runs from the energy of the fuel. There's no free lunch!
3. Windmills generate power by taking KE from the wind, so the wind is slowed by interaction with the windmill blades. So yes, it would be windier behind the windmills if they weren't there. And what is the source of wind energy? That's right—the sun.

twice the power of another does not necessarily produce twice as much work or go twice as fast as the less powerful engine. Twice the power means it will do the same amount of work in half the time or twice the work in the same time. The main advantage of a powerful automobile engine is that it can get the automobile up to a given speed in a shorter time than a less powerful engine.

Here's another way to look at power: a liter (L) of gasoline can do a given amount of work, but the power produced when we burn it can be any amount, depending on how *fast* it is burned. The liter may produce 50 units of power for a half hour in an automobile or 90,000 units of power for one second in a Boeing 747.

The unit of power is the joule per second (J/s), also known as the watt (in honor of James Watt, the eighteenth-century developer of the steam engine). One watt (W) of power is expended when one joule of work is done in one second. One kilowatt (kW) equals 1000 watts. One megawatt (MW) equals one million watts. In the United States we customarily rate engines in units of horsepower and electricity in kilowatts, but either may be used. In the metric system of units, automobiles are rated in kilowatts. (One horsepower is the same as three-fourths of a kilowatt, so an engine rated at 100 horsepower is a 75-kW engine.)

Machines

Underlying every machine is the *conservation of energy* concept. A *machine* is a device for multiplying forces or simply changing the direction of forces. Consider one of the simplest machines, the **lever** shown in the sketch. At the same time we do work on one end of the lever, the other end

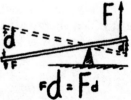

does work on the load. We see that the direction of force is changed, for if we push down, the load is lifted up. If the heat from friction forces is small enough to neglect, the work input will be equal to the work output.

Work input = work output

Since work equals force times distance, we can say;

$$(\text{force} \times \text{distance})_{\text{input}} = (\text{force} \times \text{distance})_{\text{output}}$$

If a lever's pivot point, the *fulcrum*, is exactly midway between the input force and the output force, then no multiplication of force occurs—only the direction of force is changed. But if the fulcrum is relatively close to the load, then a small input force will produce a large output force. This is because the input force is exerted through a large distance, and the load is moved over a correspondingly short distance. In this way, a lever can multiply forces. But no machine can multiply work or multiply energy. That's a conservation of energy no-no!

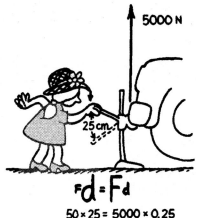

5000 N

25 cm

$$^Fd = F_d$$

$$50 \times 25 = 5000 \times 0.25$$

A child could use the principle of the lever in jacking up the front end of an automobile. By exerting a small force through a large distance, she is able to provide a large force acting through a small distance. Consider the ideal example illustrated in the sketch. Every time she pushes the jack handle down 25 centimeters, the car rises only a hundredth as far, but with 100 times the force.

A system of pulleys is a simple machine that multiplies force at the expense of distance. One can exert a relatively small force through a relatively large distance and lift a heavy load through a relatively short distance. With the ideal pulley system such as that shown to the right, the man pulls 10 meters of rope with a force of 50 newtons and lifts 500 newtons up a vertical distance of 1 meter. The energy the man expends in pulling the rope is numerically equal to the increased potential energy of the 500-newton block.

500 N

$$^Fd = F_d$$

Some physics textbooks examine the various types of levers. Type 1 levers have the fulcrum between the applied force and the load. Type 2 levers have the load between the fulcrum and the applied force, and type 3 have the applied force between the fulcrum and the load. The sketch below shows these classes of levers, but we won't discuss them further.

Any machine, lever or otherwise, that multiplies force does so at the expense of distance. Likewise, any machine that multiplies distance, such as the construction of your forearm and elbow, does so at the expense of force. So a machine can multiply force, or a machine can multiply distance moved. But get this: no machine or device can multiply *both* force and distance moved; no machine can multiply energy; no machine can put out more energy than is put into it. Machines—whether mechanical, electrical, or otherwise—don't create energy. Machines transfer energy from one location to another or transform one form of energy to another. Machines, like everything, are subject to the law of energy conservation.

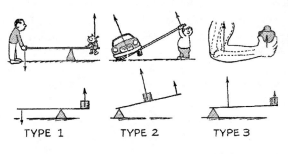

TYPE 1   TYPE 2   TYPE 3

Efficiency

Our previous examples were of ideal machines: 100 percent of the work input was transformed to work output. An ideal machine would operate at 100 percent efficiency. In practice, this doesn't happen and we can't expect it to happen. In any transformation some energy is dissipated to molecular kinetic energy—heat. This makes the machine and the environment warmer.

Even a lever rocks upon its fulcrum and converts a small fraction of the input energy into heat. We may do 100 joules of work and get out 98 joules of work. The lever is then 98 percent efficient, and we waste only 2 joules of work input on heat. If the girl jacking up the car puts in 100 joules of work and increases the potential energy of the car by 60 joules, the jack is 60 percent efficient; 40 joules of work have been used to do the work in overcoming the friction force, which appears as heat. In a pulley system, a larger fraction of input energy goes into heat. If we do 100 joules of work, the forces of friction acting through the distances through which the pulleys turn and rub about their axles may dissipate 60 joules of energy as heat. Thus, the work output is only 40 joules, and the pulley system has an efficiency of 40 percent. The lower the efficiency of a machine, the greater the amount of energy wasted as heat.

Inefficiency exists whenever energy is transformed from one form to another. **Efficiency** can be expressed by the ratio

$$\text{Efficiency} = \frac{\text{work done}}{\text{energy used}}$$

An automobile engine transforms chemical energy stored in fuel into mechanical energy. The bonds between the molecules in the petroleum fuel break up when the fuel burns by reacting with the oxygen in the air. Carbon atoms then bond with oxygen to form carbon monoxide and carbon dioxide

> **Question**
>
> Suppose a miracle car has a 100 percent efficient engine and burns fuel that has an energy content of 40 MJ/L. If the air drag and overall frictional forces on the car traveling at highway speed is 1000 N, what is the upper limit in distance per liter the car could go at this speed?

in which the bonds have less energy stored in them than in the original bonds. Some of the remaining energy goes into running the engine. We'd like all this energy converted into mechanical energy: that is, we'd like an engine that is 100 percent efficient. This is impossible because some of the energy goes out in the hot exhaust gases, and nearly half is wasted in the friction of the moving engine parts. In addition to these inefficiencies, the fuel doesn't even burn completely, and a certain amount of fuel energy goes unused.

Look at the inefficiency that accompanies transformations of energy this way: in any transformation there is a dilution of available *useful energy*. The amount of usable energy decreases with each transformation until ultimately nothing is left but thermal energy at the temperature of the surroundings. If you study thermodynamics, you'll see that thermal energy is useless for doing work unless it can be transformed to a lower temperature. Heat, diffused into the environment as thermal energy, is the graveyard of useful energy.

Kinetic Energy and Momentum Compared

Note the parallel structure of work-energy with impulse-momentum of the previous chapter. In both we see that changes in an object's motion depend on both force and how long the force acts. In the previous chapter, "how long" meant time, where we called the quantity "force × time" *impulse*—which led to the concept of *momentum*. Here we call "how long" distance, and we call the quantity "force × distance" *work* which leads to energy. Some two centuries ago, before the roles of time and distance on moving things were distinguished, physicists spoke of the *impetus* of moving objects. Not until it was

> **Answer**
>
> From the definition work = force x distance, simple rearrangement gives distance = work/force. If all 40 million J of energy in 1 L were used to do the work of overcoming the air drag and frictional forces, the distance would be:
>
> $$\text{Distance} = \frac{\text{work}}{\text{force}} = \frac{40,000,000 \text{ J} / \text{L}}{1,000 \text{ N}} = 40,000 \text{ m} / \text{L} = 40 \text{ km} / \text{L}$$
>
> If we streamline the car to cut air drag and reduce friction by half, to 500 N, then we can get 80 km/L. The important point here is that even with a perfect engine, there is an upper limit of fuel economy dictated by the conservation of energy.

generally realized that force × time and force × distance had different consequences, did *impetus* give way to the separate concepts of momentum and kinetic energy.

If momentum and kinetic energy are new concepts in your thinking, it will be enough if you can see the connections between force and a body's changes in motion. Distinguishing between momentum and kinetic energy, on the other hand, is a deeper undertaking. (Skimming or skipping ahead to the next section, **Energy Sources**, at this point may be a good idea for readers wanting light coverage.)

Momentum, like velocity, is a vector quantity and is therefore directional and subject to cancellation. But kinetic energy is a non-vector (*scalar*) quantity, like mass, and can never be canceled. The momenta of two cars just before a head-on collision may cancel to zero, and the combined wreck after collision will have the same zero value for the momentum—but the energies add, as evidenced by the deformation and heat after collision. For example, the momenta of two firecrackers approaching each other may cancel, but when they explode, there is no way their energies can cancel. Energies transform to other forms; momenta do not. The vector nature of momentum is a key difference from the scalar nature of kinetic energy.*

Another difference is the velocity dependence of the two. Whereas momentum depends on velocity (mv), kinetic energy depends on the square of velocity ($1/2mv^2$).** An object that moves with twice the velocity of another object of the same mass has twice the momentum but four times the kinetic energy. So when a car going twice as fast crashes, it crashes with four times the energy.

An example that neatly illustrates the distinction between momentum and kinetic energy is the firing of a gun. Recall that the momentum of the emerging bullet is equal and opposite to the momentum of the recoiling gun (neglecting gunpowder gases). But the KE (kinetic energy) of the bullet is enormously larger than the KE of the recoiling gun. For example, if the bullet's speed is a hundred times greater than the gun's recoil speed (which means the bullet must have one-hundredth the mass of the gun) then the KE of the bullet is a hundred times greater than the KE of the gun.***

---

*   If you're into mathematics, you may know that when a vector quantity (velocity) is multiplied by a scalar quantity (mass), the product is also a vector (momentum). But a vector quantity squared (velocity ^{2}) is a scalar. So KE is a scalar quantity. It can't be canceled. Although momenta can be combined in such a way to cancel to zero, there is no way to combine KEs to equal zero.

**   Interestingly enough, kinetic energy is equal to the momentum squared divided by twice the mass. To see this, let momentum mv be p. Then KE = $1/2 \, mv^2$ = $1/2 \, mvv$ = $1/2 \, pv$. Multiply by m/m, and we see KE = $pmv/2m = p^2/2m$.

***Let m be the mass of the bullet, and $100m$ the mass of the gun. Then $KE_{gun} = 1/2 \, (100m)v^2 = 50mv^2$. By momentum conservation, if the speed of the gun is v, the speed of the bullet is $100v$. $KE_{bullet} = 1/2 \, m(100v)^2 = 1/2 \, m(10000)v^2 = 5000 \, mv^2$. So the bullet's KE is 100 times greater than the recoiling gun's KE.

Another example that nicely distinguishes between momentum and kinetic energy is the novel
swinging balls device—the "swinging wonder. " When
a single ball is raised and allowed to swing into the
array of other identical balls, a single ball from the other
side pops out. When two balls are similarly raised and
released, presto—two balls on the other side pop out.
The number of balls impinging on the array is always
the same as the number of balls that emerge. Clearly,
momentum before equals momentum after. That is, $mv = mv$, or $2mv = 2mv$,
and so on. The intriguing question arises: when a single ball is raised, released, and makes impact, why cannot two balls emerge with half the speed?
Or if two balls make impact, why cannot one ball emerge with twice the speed?
If either of these cases occurred, the momentum before would still be equal to
the momentum after: $mv = 2m(1/2\ v)$; or $2mv = m(2v)$. Intriguingly, this never
happens—nor can it happen. Why? Because momentum is not the only quantity that is conserved. Since the collisions are quite elastic, with very little
energy transforming to heat and sound, to a good approximation the kinetic
energy before equals the kinetic energy after. That is, $1/2\ mv^2_{before} = 1/2\ mv^2_{after}$
Consider dropping two balls with one emerging at twice the speed. Then will
$1/2\ (2m)v^2 = 1/2\ m(2v)^2$? The answer is no! If this case were to occur, there
would be more energy after the collision than before (I'll leave it to you to
figure how much more). Give this some thought and you'll see there is a reason why, for identical balls, the number of balls that make impact will always
equal the number of balls that emerge.

Why is this device called the "swinging wonder?" Because the unequal-
number-of-balls situation and its impossibility have left many people wondering—and wondering—and wondering. Momentum and kinetic energy are properties of moving things, but we see they are different from each other. If this
distinction is not really clear to you, you're in good company. Failure to make
this distinction, when impetus was in vogue, resulted in disagreements and
arguments between the best British and French physicists for two centuries.

Energy Sources

Except for nuclear power, the source of practically all of our energy is the sun.
This includes the energy we obtain from the combustion of petroleum, coal,
natural gas, and wood, for these materials are the result of photosynthesis, a
biological process that incorporates the sun's radiant energy. There are many
other ways of using the sun as a source of energy. Sunlight, for example, can
be directly transformed into electricity by way of photo-voltaic cells, like those
found in solar-powered calculators. Solar radiation can also be used indirectly
to generate electricity. Sunlight evaporates water, which later falls as rain;
rainwater flows into rivers and turns generator turbines as it returns to the
sea. Using mirrors, solar radiation can be concentrated to heat water into steam,

which can also be used to turn generator turbines. Furthermore, wind is solar energy that has already been converted into mechanical energy. The mechanical energy of wind can be used to turn generator turbines within specially equipped windmills. A problem with these sources of power is gathering it, for the energy is dilute. Concentrated energy sources such as fossil fuels and uranium are the choice contenders for large-scale power production.

The most concentrated form of available energy is in uranium and pluto-nium-nuclear fuels. Nuclear fission power plants account for nearly 80 percent of electrical power in France, and usage is growing. The French recycle and monitor radioactive byproducts. In the United States, some 20 percent of the electricity is presently produced by nuclear power plants, and this percentage is scheduled to diminish in coming years. While public phobia about nuclear power is relatively absent in France, in the United States it is more entrenched than the similar phobia about electricity in the 1800s. Americans see radioactive byproducts as poisonous wastes to be buried for all time. Public fear is fueled by the misconception that radioactivity is the product of technology, when in fact more radioactivity emanates from the ground than from man-made sources, including nuclear power plants. Bathe in a natural hot spring or observe a natural geyser, and you're witnessing a byproduct of radioactivity, for radioactivity is the primary source of heat in the earth's interior. Radiation is a part of nature that was present, even in stronger quantities, before the advent of man, and will persist for all time. But if you mention in public the `R word' or the `N word' in a positive tone, you risk being seen as an ally of Darth Vader. Never mind that we're squandering the earth's supply of fossil fuels at ever increasing rates. Recall how the peoples of Mexico and Central America were astonished to see that the conquering Spaniards valued their works of art only for their yellow-metal content. Perhaps our great grandchildren will be as astonished that we similarly value the earth's supply of complex hydrocarbons as a source of heat—hydrocarbons synthesized over many millennia. In the coming century, perhaps clearer thinking will prevail.

Geothermal energy is an offspring of radioactive heating, which converts water to steam for generating electricity. Geothermal energy is predominantly limited to areas of volcanic activity, such as Iceland, New Zealand, Japan, and Hawaii. Another method that holds promise is dry-rock geothermal power, where cavities are made in deep dry hot rock into which water is introduced. When the water turns to steam it is piped to a turbine at the surface. Then it is returned to the cavity for re-use.

Nuclear and geothermal power have an added advantage of not polluting the environment. The combustion of fossil fuels, on the other hand, which is as American as Mom's apple pie, leads to increased atmospheric concentrations of carbon dioxide, sulfur dioxide, and many other pollutants.

Another non-polluting source of energy, which was successfully demonstrated off the Kona Coast in Hawaii, is OTEC, the acronym for Ocean Thermal Energy Conversion. (Due to recent Federal cutbacks, this program is no

longer funded.) Like wind power and hydropower, OTEC power comes ultimately from sunlight and needs no other fuel. It is a low-temperature thermal
engine that runs on the temperature difference between solar-lit 26°C tropical
surface waters and deep, dark 4°C waters. This relatively low temperature
difference is enough to run its thermal engine, although at low efficiency (the
greater the temperature difference, the greater the efficiency, as we shall see
in Book Two). The low efficiency of OTEC is compensated by circulating large
amounts of water. Cold deep waters are circulated to the surface, where they
produce an additional bonus—the addition of nitrogen and other nutrients
that have been locked away from the food chain in darkness for centuries.
This upwelling of nutrient-rich water exposed to sunlight should produce an
explosion of phytoplankton growth comparable to that obtained when fertil-
izers are added to land crops. So the promise of OTEC is not only pollution-
free electric energy, but increased food from the sea.

As the world population increases, so does our need for both energy
and food. Common sense dictates that as new sources are being developed,
we should continue to optimize present sources and use what we consume
efficiently and wisely.

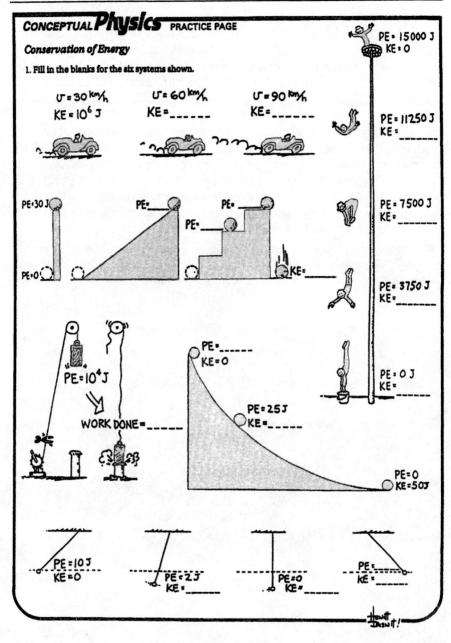

CONCEPTUAL **Physics** PRACTICE PAGE

Conservation of Energy

1. Fill in the blanks for the six systems shown.

$v = 30$ km/h
KE = 10^6 J

$v = 60$ km/h
KE = _____

$v = 90$ km/h
KE = _____

PE = 15000 J
KE = 0

PE = 11250 J
KE = _____

PE·30 J

PE = _____

PE = _____

PE = _____

PE·0

KE = _____

PE = 7500 J
KE = _____

PE = 3750 J
KE = _____

PE = 10^4 J

WORK DONE = _____

PE = _____
KE = 0

PE = 25 J
KE = _____

PE = 0 J
KE = _____

PE = 0
KE = 50 J

PE = 10 J
KE = 0

PE = 25 J
KE = _____

PE = 0
KE = _____

PE = _____
KE = _____

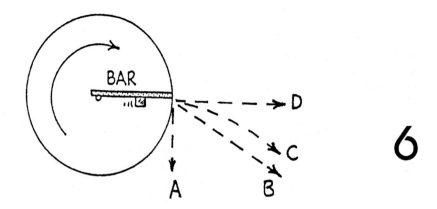

Rotational Motion

One of the intriguing problems that Burl Grey posed to me years ago on the sign-painting stages in Miami was what path would an object take that slides off a rotating turntable? Burl considered a record player (remember them?). Place a pebble or some small object on it close to the center and start it up. The record rotates and the object slides off the edge. We pondered the path it might take—straight line radially outward? A straight line tangential to the edge? A straight line at an angle between the two? Or a curved line? What path? To see this better, we supposed a bar or some straightedge were affixed to the record, so the sliding object would slide along it in a radial direction before leaving the edge. I resolved that during that evening I would do the demonstration in my apartment and see the answer. I'd sprinkle some talcum powder on the floor so the sliding object would leave a tell-tale path. And what did I do when I got home after work? I was distracted and didn't do the experiment. The next day, Burl and I got into other topics, and the record problem went unsolved. It wasn't until I went to prep school the following year that I resolved the problem. I was studying vectors, and whamo! The solution became evident. I tell this story when I teach vectors and rotational motion. I tell how the answer to the problem remained a secret for more than a year before I got it. Then after posing the problem and having my students check their hypotheses with their neighbors, I ask if they'd like my answer. Yes, is always the reply. Then I tell them this was a secret for me, and if I'm to tell them, I first want to know if they can keep a secret. Yes, they say, to which I respond, "So can I." I don't tell them the answer, for the same reason that I'm glad Burl didn't tell me. Good teaching isn't providing answers—it's asking good questions.

Learning about the physics of rotational motion means placing the academic plow a bit deeper. If previously covered material has had time to jell in students' minds, then success is assured. This is because the concepts of rotational motion are the rotational counterparts of the concepts of inertia, force, speed, and momentum. We speak of rotational inertia, torque, rotational speed, and rotational momentum. We'll see how rotational inertia affects rotation, how forces applied in a certain way produce torques that tend to rotate things, and how linear momentum is extended to angular or rotational momentum. This is interesting stuff, especially in a classroom with a teacher who likes to show demonstrations. Rotational motion allows for a wide variety of interesting demos.

Rotational Inertia

Just as an object at rest tends to stay at rest, and an object in motion tends to remain moving in a straight line, *an object rotating about an axis tends to remain rotating about the same axis unless interfered with by some external influence.* (We shall see shortly that this external influence is properly called a *torque.*) The property of an object to resist changes in rotation is called **rotational inertia.**[*] Things that rotate tend to remain rotating, while non-rotating things tend to remain non-rotating. In the absence of outside influences, a rotating top keeps rotating, while a top at rest stays at rest.

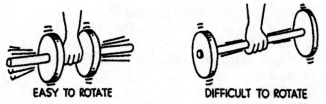

EASY TO ROTATE   **DIFFICULT TO ROTATE**

Like inertia for linear motion, the rotational inertia of an object also depends on its mass. The stone disk that rotates beneath a potter's wheel is very massive, and once spinning, tends to remain spinning. But unlike linear motion, rotational inertia depends on the distribution of the mass with respect to the axis of rotation. The greater the distance between the bulk of an object's mass and its axis of rotation, the greater the rotational inertia. This is evident in industrial flywheels that are constructed so that most of their mass is concentrated far from the axis, along the rim. Once rotating, they have a greater tendency to remain rotating. When at rest, they are more difficult to get rotating. The greater the rotational inertia of an object, the harder it is to change the rotational state of that object. This fact is employed by a circus tightrope walker who carries a long pole to aid balance. Much of the mass of the pole is far from the axis of rotation, its midpoint. The pole therefore has considerable rotational inertia. If the tightrope walker starts to topple over, a tight grip on the pole rotates the pole. But the rotational inertia of the pole resists, giving

---

* Often called *moment of inertia.*

the tightrope walker time to readjust his or her balance. The longer the pole, the better. And better still if massive objects are fixed to the ends. But a tightrope walker with no pole can at least extend the arms full length to increase the body's rotational inertia.

The rotational inertia of the pole, or any object, depends on the axis around which it rotates.* Compare the different rotations of a pencil. Consider three axes—one, through its central core parallel to the length of the pencil, where the lead is; two, through the perpendicular midpoint axis; and three, an axis perpendicular to one end. Rotational inertia is very small around the first position, for most all the mass is very close to the axis. It's easy to rotate the pencil to and fro along its long axis between your finger tips. About the second axis, like that used by the tightrope walker in the illustration above, rotational inertia is greater. About the third axis, at the end of the pencil so it swings like a pendulum, rotational inertia is greater still.

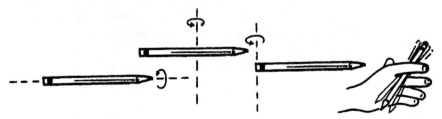

A long pendulum has a greater rotational inertia than a short pendulum and, therefore, swings back and forth more slowly than a short one. When we walk, we allow our legs to swing with the help of gravity, that is, at a pendulum rate. Just as a long pendulum takes a long time to swing to and fro, a long-legged person tends to walk with slower strides than a person with short

---

* When the mass of an object is concentrated totally at one distance, the radius r from the axis of rotation, like a simple pendulum bob or a thin ring, rotational inertia I is equal to the mass m multiplied by the square of the radial distance. For this special case, $I = mr^2$.

legs. Running with your legs straight is difficult, which is why you bend your legs when running—you reduce their rotational inertia so you can rotate them back and forth more quickly. The different strides of creatures with different leg lengths is especially evident in animals: giraffes, horses, and ostriches run with a slower gait than dachshunds, mice, and bugs.

Due to rotational inertia, a solid cylinder starting from rest will roll down an incline faster than a ring or hoop. All rotate about a central axis, and the shape that has most of its mass far from its axis is the ring. So for its weight, a ring has more rotational inertia and is harder to start rolling. Any solid cylinder will outroll any ring on the same incline. This doesn't seem plausible at first, but remember that any two objects, regardless of mass, will fall together when dropped. They will also slide together when released on an inclined plane. When rotation is introduced, the object with the larger rotational iner-

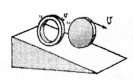

tia *compared to its own mass* has the greater resistance to a change in its motion. Hence, any solid cylinder will roll down any incline with more acceleration than any hollow cylinder, regardless of mass or radius. A hollow cylinder has more "laziness per mass" than a solid cylinder. Try it and see!

A homework exercise I almost always assign is to compare the rotational rates of a can of soda and a can of solid food. When both roll down an incline, the one that wins is the one with the least rotational inertia relative to its mass. Points are given for the correct answer—the can of soda. And bonus points are given for a plausible explanation. It so happens that the soda barely rolls as the can rolls down an incline. Like the soup that stays put when you turn the bowl, the soda simply slides without

1. Consider standing and balancing a hammer upright on the tip of your finger. If the head of the hammer is heavy and the handle long, would it be easier to balance with the end of the handle on your fingertip so that the head is at the top, or the other way around with the head at your fingertip and the end of the handle at the top?

2. Consider a pair of metersticks standing nearly upright against a wall. If you release them, they'll rotate to the floor in the same time. But what if one has a massive hunk of clay stuck to its top end? Will it rotate to the floor in a longer or shorter time?

3. Just for fun, and since we're discussing round things, why are manhole covers round in shape?

rotating inside the rolling can. The soda increases the mass of the can, but not its rotational inertia. The can with the solid contents, however, has more rotational inertia because the contents are forced to roll with the can. After this is discussed and shown in class, I follow up with a test question asking which will roll down an incline faster—a can of water or a can of solid ice.

Torque

Hold the end of a meterstick horizontally with your hand. Dangle a weight from it near your hand and you can feel the stick twist. Now slide the weight farther from your hand and the twist is more. But the weight is the same. The force acting on your hand is the same. What's different is the *torque*.

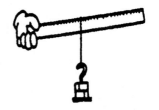

1. Stand the hammer with the handle at your fingertip and the head at the top. Why? Because it will have more rotational inertia this way and be more resistant to a rotational change. Those acrobats you see on stage who balance their friends at the top of a long pole have an easier task when their friends are at the top of the pole. A pole empty at the top has less rotational inertia and is more difficult to balance!

2. Try it and see! (If you don't have clay, fashion something equivalent.)

3. Not so fast on this one. Give it some thought if you haven't come up with an answer. Then look to the end of the chapter for an answer.

A torque (rhymes with *dork*) is the rotational counterpart of force. Force tends to change the motion of things; torque tends to twist or change the state of rotation of things. If you want to make a stationary object move, apply force. If you want to make a stationary object rotate, apply torque. Torque differs from force just as rotational inertia differs from ordinary inertia: both involve distance from the axis of rotation. In the case of torque, this distance is called the lever arm. We define *torque* as the product of this lever arm and the force that tends to produce rotation around this particular distance:

Torque = lever arm × force

Torques are intuitively familiar to youngsters playing on a seesaw. Kids can balance a seesaw even when their weights are unequal. Weight alone doesn't produce rotation. Torque does, and kids soon learn that the distance they sit from the pivot point is every bit as important as weight. The torque produced by the boy on the right tends to produce clockwise rotation, while

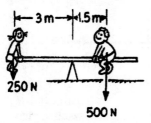

torque produced by the girl on the left tends to produce counterclockwise rotation. If the torques are equal, making the net torque zero, no rotation is produced. (Interestingly, the terms clockwise and counterclockwise still have meaning. But as more and more digital clock faces replace those with hands, these terms, like so many others, will no longer be as meaningful.)

Suppose that the seesaw is arranged so that the half-as-heavy girl is suspended from a 4-meter rope hanging from her end of the seesaw. She is now 5 meters from the fulcrum, and the see-saw is still balanced. We see that the lever-arm distance is still 3 meters and not 5 meters. The lever arm about any axis of rotation is the perpendicular distance from the axis to the line along which the force acts. This will always be the shortest distance between the axis of rotation and the line along which the force acts.

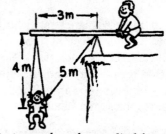

This is why a stubborn bolt is more likely to turn when the applied force is perpendicular to the wrench handle, rather than at an oblique angle. Note in the left part of the sketch that the lever arm is shown by the dotted line and is less than the length of the wrench handle. In the middle part, the lever arm is equal to the length of the wrench handle. In the right side we see the lever arm is extended with a pipe to produce a greater torque.

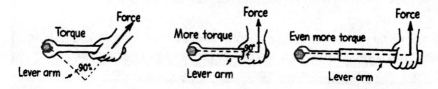

Questions

1. If a pipe effectively extends a wrench handle to three times its length, by how much will the torque increase for the same applied force?

2. Helen Yan and Dan Johnson are balanced on the seesaw. Suppose Helen suddenly gains 50 N, as if she were handed a bag of apples. Where should she sit in order to balance, assuming heavier Dan does not move?

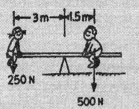

Center of Mass and Center of Gravity

Throw a baseball into the air, and it will follow a smooth trajectory. Throw a baseball bat spinning into the air, and its path is not smooth; its motion is wobbly, and it seems to rotate all over the place. But, in fact, it rotates about a very special place, a point called the center of mass.

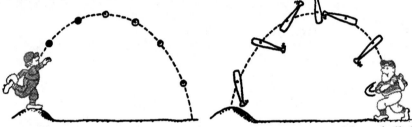

For a given body, the center of mass is the average position of all the mass that makes up the object. For example, a symmetrical object like a ball can be thought of as having all its mass concentrated at its geometric center; by contrast, an irregularly shaped object such as a baseball bat has more of its mass toward one end. The center of mass of a baseball bat, therefore, is toward the thicker end. A solid cone has its center of mass exactly one-fourth of the way up from its base.

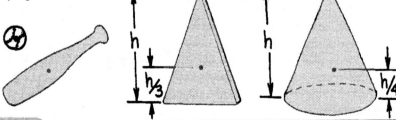

Answers

1. Three times. (This method of increasing torque sometimes results in shearing off the bolt!)

2. Helen should sit 1/2 m closer to the center. Then her lever arm is 2.5 m. This checks:

$$300 \text{ N} \times 2.5 \text{ m} = 500 \text{ N} \times 1.5 \text{ m}$$

Center of gravity is a term popularly used to express center of mass. The center of gravity is simply the average position of weight distribution. Since weight and mass are proportional, center of gravity and center of mass refer to the same point of an object.* The physicist prefers to use the term *center of mass*, for an object has a center of mass whether or not it is under the influence of gravity. However, we shall use either term to express this concept and favor the term *center of gravity* , or simply CG for short, when consideration of weight is involved.

The multiple-flash photograph shows a top view of a wrench sliding across a smooth horizontal surface. Note that its CG, indicated by the dark mark, follows a straight-line path, while other parts of the wrench wobble as they move across the surface. Since there is no net force acting on the wrench, its CG moves equal distances in equal time intervals. The motion of the spinning wrench is the combination of the straight-line motion of its CG and the rotational motion about its CG.

If the wrench were instead tossed into the air, no matter how it rotates, its CG would follow a smooth parabolic path. The same is true for an exploding cannonball. The internal forces that act in the explosion do not change the CG of the projectile. Interestingly enough, if air resistance can be neglected, the CG of the dispersed fragments as they fly though the air will be no different than if the explosion didn't occur. The CG of the dispersed fragments follows the path that the non-exploded cannonball would follow.

Locating the Center of Gravity

The CG of a uniform object such as a meter stick is at its midpoint, for the stick acts as though its entire weight were concentrated there. Support at that single point supports the whole stick. Balancing an object provides a simple method of locating its CG. The many small vectors in the sketch of the meter stick rep-

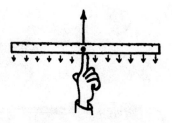

---

* There can be a small difference between center of gravity and center of mass when an object is large enough for gravity to vary from one part to another. For example, the center of gravity of the Sears Tower is about 1 millimeter below its center of mass. This is due to the lower stories being pulled a little more strongly by the earth's gravity than the upper stories. For everyday objects (including tall buildings!) we can use the terms center of gravity and center of mass interchangeably.

resent the pull of gravity all along the stick. All these can be combined into a resultant force acting through the CG. The entire weight of the stick in effect acts at this single point. That's why we can balance the stick by applying a single upward force in a direction that passes through this point.

The CG of any freely suspended object lies directly beneath (or at) the point of suspension. If a vertical line is drawn through the point of suspension, the CG lies somewhere along that line. To determine exactly where it lies along the line, we have only to suspend the object from some other point and draw a second vertical line through that point of suspension. The CG lies where the two lines intersect.

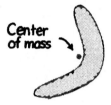

Center of mass

The CG of an object may be a point where no mass exists. For example, the center of mass of a ring or a hollow sphere is at the geometrical center where no matter exists. Similarly, the center of mass and CG of a boomerang is outside the physical structure, not within the material making up the boomerang.

Questions

1. Where is the center of mass of the earth's atmosphere?

2. Why is it dangerous to slide open the top drawers of a fully loaded file cabinet that is not secured to the floor?

3. When a car drives off a cliff, why does it rotate forward as it falls?

Answers

1. Like a giant basketball, the earth's atmosphere is a spherical shell with its center of mass at the earth's center.

2. The filing cabinet is in danger of tipping because the CG may extend beyond the support base (see next page).

3. When all wheels are on the ground, the car's CG is within a support base. But when the car drives off a cliff, the front wheels are first to leave the ground, and the support base shrinks to the area between the rear wheels. So the car's CG extends beyond the support base and it rotates, as would the Leaning Tower of Pisa if its CG extended beyond its support base (again, next page).

Stability

The location of the CG is important for stability. If we drop a line straight down from the CG of an object of any shape and it falls inside the base of the object, it is in stable *equilibrium*—it will balance. If it falls outside the base, it is unstable. Why doesn't the famous Leaning Tower of Pisa topple over? As we can see in the sketch, a line from the CG of the tower falls inside its base, so the Leaning Tower has stood for several centuries.*

To reduce the likelihood of tipping, it is usually advisable to design objects with a wide base and low center of gravity. The wider the base, the higher the CG must be raised before the object will tip over. More energy must be expended in raising a CG higher.

When you stand erect (or lie flat), your CG is within your body. Why is the CG lower in an average woman than in an average man of the same height? The sketch shows two vertical baseball bats, one standing on its heavier end and the other standing on its lighter handle end. Can you see how this relates to the relative CGs of men and women? In both cases, the CG is within the bats or within the bodies of people when standing erect. What happens to your CG when you bend over?

If you are in reasonably good shape, you can bend over and touch your toes without bending your knees—providing you are not standing with your back against a wall. Ordinarily, when you bend over and touch your toes, you extend your behind so that your CG is above a point of support—your

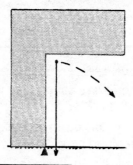

---

* To combat the gradually increasing leaning, 400 tons of lead ingots were stacked on a newly reinforced base of the Leaning Tower in 1993 to reverse the 800 years of slow tilting, and to pull the 187-foot pillar of white marble back toward the vertical. The change is less than 1/6 inch from the 16 feet by which the top overhangs the base—cause for celebration. Although the stack of lead ingots are somewhat of an eyesore, they are a step to some more elegant solution. The hope is to restore the tower from a 10° tilt to a 9° tilt and fix it there. Nobody in Pisa's tourist industry wants it exactly vertical.

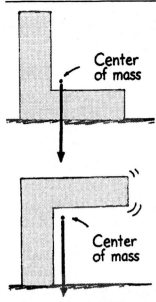

Center of mass

Center of mass

feet. If you attempt to do this when standing against a wall, however, you cannot extend your behind to counterbalance yourself, and your CG soon protrudes beyond your feet as at the left in the sketch. You are off-balance and you rotate.

You rotate because of an unbalanced torque. This is evident in the two L-shaped objects shown. The top one is in stable equilibrium, and the bottom one is unstable. The CG extends beyond the support base. A resulting torque causes it to topple clockwise.

Try balancing the pole end of a broom or mop on the palm of your hand. The support base is quite small and relatively far beneath the CG, so it's difficult. After some practice you can do it if you learn to make slight movements of your hand to respond exactly to variations in balance. You learn to avoid under-responding or over-responding to the slightest variations in balance. Similarly, high-speed computers help massive rockets remain upright when they are launched. Variations in balance are quickly sensed. The computers regulate the firings at multiple nozzles to make corrective adjustments in a way quite similar to the way your brain coordinates your adjustive action when balancing a long pole on the palm of your hand. Both feats are truly amazing.

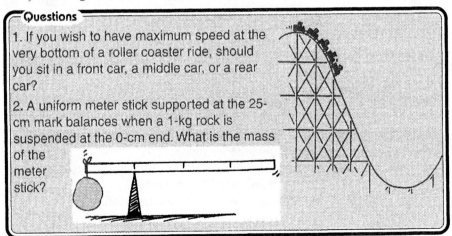

Questions

1. If you wish to have maximum speed at the very bottom of a roller coaster ride, should you sit in a front car, a middle car, or a rear car?

2. A uniform meter stick supported at the 25-cm mark balances when a 1-kg rock is suspended at the 0-cm end. What is the mass of the meter stick?

Angular Momentum

Rotating things, whether a colony in space, a cylinder rolling down an incline, or an acrobat doing a somersault, keep on rotating until something stops rotation. A rotating object has an "inertia of rotation." Recall from Chapter Four that all moving objects have "inertia of motion" or momentum—the product of mass and velocity. We call this kind of momentum **linear momentum**. Similarly, the "inertia of rotation" of rotating objects is called **angular momentum**.

A planet orbiting the sun, a rock whirling at the end of a string, and the tiny electrons whirling about atomic nuclei all have angular momentum.*

Angular momentum is defined as the product of rotational inertia and rotational velocity.

Angular momentum = rotational inertia × rotational velocity

It is the counterpart of linear momentum:

Linear momentum = mass × velocity

Like linear momentum, angular momentum is a vector quantity, having direction as well as magnitude. In this book, we won't treat the vector nature of momentum (or even of torque, which also is a vector), except to acknowledge the remarkable action of the gyroscope. The rotating bicycle wheel in the photo on the facing page shows what happens when a torque caused by earth's gravity acts to change the direction of its angular momentum (which is along the wheel's axle). That's my late son James holding the wheel. He was standing on a turntable, and the pull of gravity that acts to topple the wheel over and change its rotational axis instead causes it to precess (move sideways) in a circular path about a vertical axis. So as the wheel spins, the

---

Answers

1. Maximum speed occurs when KE is maximum, which occurs when PE is minimum—when the CG of the cars is lowest. So sit in a middle car, nearest the CG of the coaster, and when you're at the bottom the cars will have maximum speed.

2. The mass of the meter stick is 1 kg. Why? The system is in equilibrium, so any torques must be balanced: the torque that results from the weight of the rock is balanced by the equal but oppositely directed torque that results from the weight of the stick at its CG, the 50-cm mark. The support force at the 25-cm mark is applied midway between the rock and the CG of the stick, so the lever arms about the support point are equal (25 cm). This means that the weights and, hence, the masses of the rock and stick must also be equal. Interestingly enough, the CG of the *rock and stick combination* is at the 25-cm mark—directly above the fulcrum. (Note that we don't have to go through the laborious task of considering the fractional parts of the stick's weight on either side of the fulcrum, for the CG of the whole stick really is at one point—the 50-cm mark!

---

* Angular momentum is a vector quantity: it has direction and magnitude. When a direction is assigned to rotational speed, we call it *rotational velocity* (often called *angular velocity*). Rotational velocity is a vector whose magnitude is the rotational speed. By convention, the rotational velocity vector and the angular momentum vector have the same direction and lie along the axis of rotation. In follow-up reading you may learn that rotational velocity is measured in units of radians per second (rad/s), where 0.1 rad/s is approximately 1 RPM.

turntable rotates. You must do this yourself to believe it fully. You probably won't fully understand it until a later time.

For the case of an object that is small compared to the radial distance of its axis of rotation, like a stone swinging from a long string or a planet orbiting around the sun, the angular momentum can be equivalently and more simply expressed as the magnitude of its linear momentum, mv, multiplied by the radial distance, r.

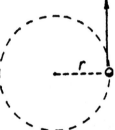

In shorthand notation,

Angular momentum = mvr

Just as an external net force is required to change the linear momentum of an object, an external net torque is required to change the angular momentum of an object. We restate Newton's first law of inertia for rotating systems in terms of angular momentum:

An object or system of objects will maintain its angular momentum unless acted upon by an unbalanced external torque.

We all know that it is easier to balance on a bicycle that is moving than on a bicycle at rest. Rotating wheels have angular momentum and are harder to tip. To tip the wheels means a change in angular momentum, which requires a greater torque than tipping wheels at rest.

Conservation of Angular Momentum

Just as the linear momentum of any system is conserved if no net forces act on the system, angular momentum is conserved for systems in rotation. If no unbalanced external torque acts on the rotating system, the angular momentum of that system is constant. This means that the product of rotational inertia and rotational velocity at one time will be the same as at any other time.

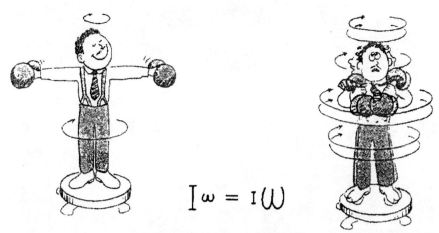

$$I \omega = I \omega$$

An interesting example illustrating angular momentum conservation is shown by the man standing on a low-friction turntable with weights extended. His rotational inertia I, with the help of the extended weights, is relatively large in this position. As he slowly turns, his angular momentum is the product of his rotational inertia and rotational velocity ω. When he pulls the weights inward, the rotational inertia of his body and the weights is considerably reduced. What is the result? His rotational speed increases! This example is best appreciated by the turning person who feels changes in rotational speed that seem to be mysterious. But it's straight physics! This procedure is used by a figure skater who starts to whirl with her arms and perhaps a leg extended and then draws her arms and leg in to obtain a greater rotational speed. Whenever a rotating body contracts, its rotational speed increases.

Similarly, when a gymnast is spinning freely in the absence of unbalanced torques on the body, angular momentum does not change. The gymnast can change rotational speed, however, by simply making variations in rotational inertia. This is accomplished by moving some parts of the body toward or away from the axis of rotation.

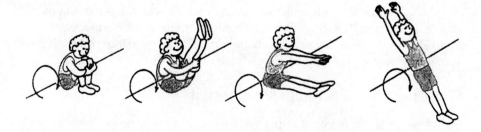

If a cat is held upside down and dropped, it is able to execute a twist and land upright even if it has no initial angular momentum. Zero-angular-momentum twists and turns are performed by turning one part of the body against the other. While falling, the cat rearranges its limbs and tail to change its rotational inertia. Repeated reorientations of the body configuration result in the head and tail rotating one way and the feet the other, so that the feet are downward when the cat strikes the ground. During this maneuver the total angular momentum remains zero. When it is over, the cat is at rest. This maneuver rotates the body through an angle, but it does not create continuing rotation. To do so would violate angular momentum conservation.

Humans can perform similar twists without difficulty, though not as fast as a cat. Astronauts have learned to make zero-angular-momentum rotations as they orient their bodies in preferred directions when floating freely in space. The law of momentum conservation is seen in the planetary motions and in the shape of the galaxies. This law will be a fact of everyday life to inhabitants of rotating space habitats who will head for these distant places.

> **Question**
> Our galaxy may have begun as a huge cloud of gas and particles. Suppose the original cloud was far larger than the present size of the galaxy, was more or less spherical, and was rotating very much more slowly than at present. Gravitation between particles would have pulled them closer. What would be the role of angular momentum conservation on the galaxy's shape and present rotational speed?

It is fascinating to note that the conservation of momentum tells us that the moon is getting farther away from the earth. This is because the earth's daily rotation is slowly decreasing due to the friction of ocean waters at the ocean bottom, just as an automobile's wheels slow down when brakes are applied. This decrease in the earth's angular momentum is accompanied by an equal increase in the angular momentum of the moon in its orbital motion around the earth. This increase in the moon's angular momentum results in the moon's increasing distance from the earth and a decrease in its speed. This increase of distance amounts to one-quarter of a centimeter per rotation. Have you noticed that the moon is getting farther away from us lately? Well, it is. Each time we see another full moon, it is one-quarter of a centimeter farther away!

> **Answer**
>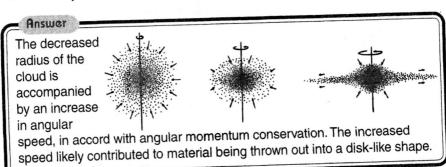
> The decreased radius of the cloud is accompanied by an increase in angular speed, in accord with angular momentum conservation. The increased speed likely contributed to material being thrown out into a disk-like shape.

Oh yes, before we progress further in this chapter, we'll give an answer to Question 3 back on page 101. Manhole covers are round because a round cover is the only shape that can't fall into the hole. A square cover, for example, can be tilted vertically and turned so it can drop diagonally into the hole. Likewise for every other shape. If you're working in a manhole and some fresh kids are horsing around above, you'll be glad the cover is round!

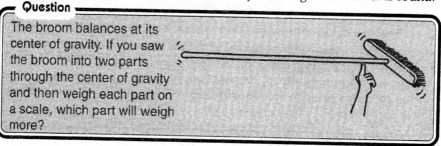

Question

The broom balances at its center of gravity. If you saw the broom into two parts through the center of gravity and then weigh each part on a scale, which part will weigh more?

Railroad Train Wheels

Why does a moving railroad train stay on the tracks? Most people assume the wheel flanges keep the wheels from rolling off. But if you look at these flanges you'll note they may be rusty. They seldom touch the track, except when they follow slots that switch the train from one set of tracks to another. So how do the wheels of a train stay on the tracks? They stay on the track because their rims are slightly tapered.

If you roll a tapered cup across a surface, it makes a curved path. The larger-diameter end rolls a greater distance per revolution and has a greater linear speed than the smaller end. If you fasten a pair of cups together at their wide ends (simply taping them together) and roll the pair along a pair of parallel tracks, the cups will remain on the track and center themselves whenever they roll off center. This occurs because when the pair rolls to the left of center, the wide part of the left cup rides on the left track while the narrow part of the right cup rides on the right track. This steers the pair toward the

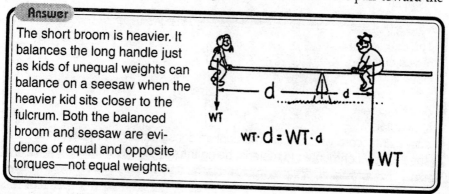

Answer

The short broom is heavier. It balances the long handle just as kids of unequal weights can balance on a seesaw when the heavier kid sits closer to the fulcrum. Both the balanced broom and seesaw are evidence of equal and opposite torques—not equal weights.

$$wt \cdot d = WT \cdot d$$

center. If it "overshoots" toward the right, the process repeats, this time toward the left, as the wheels tend to center themselves. Likewise for a railroad train, where passengers feel the train swaying as these corrective actions occur.

This tapered shape is essential on the curves of railroad tracks. On any curve, the distance along the outer part is longer than the distance along the inner part. So whenever a vehicle follows a curve, its outer wheels travel faster than its inner wheels. For an automobile this is no problem because the wheels are freewheeling, and roll independently of each other. For a train, however, like the pair of fastened cups, pairs of wheels are firmly connected so that they rotate together. Opposite wheels have the same RPM at any time. But due to the slightly tapered rim of the wheel, its speed along the track depends on whether it rides on the narrow part of the rim or the wide part. On the wide part it travels faster. So when a train rounds a curve, wheels on the outer track ride on the wide part of the tapered rims while opposite wheels ride on their narrow parts. In this way, the wheels have different linear speeds for the same rotational speed. Can you see that if the wheels were not tapered, scraping would occur and the wheels would squeal when a train rounds a curve?

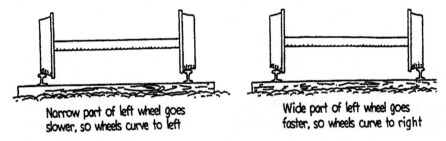

Narrow part of left wheel goes slower, so wheels curve to left

Wide part of left wheel goes faster, so wheels curve to right

Some Rotational Motion Activities You Can Do

1. Fasten a fork, spoon, and wooden match together as shown. The combination will balance nicely—on the edge of a glass, for example. This happens because the center of gravity actually "hangs" below the point of support.

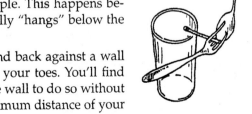

2. Stand with your heels and back against a wall and try to bend over and touch your toes. You'll find you have to stand away from the wall to do so without toppling over. Compare the minimum distance of your heels from the wall with that of a friend of the opposite sex. Who can touch their toes with their heels nearer to the wall—males or females? On the average and in proportion to height, which sex has the lower center of gravity?

3. Ask a friend to stand facing a wall. With toes against the wall, ask your friend to stand on the balls of her feet without toppling backward. Your friend won't be able to do it. Now you explain why it can't be done.

4. Rest a meter stick on two fingers as shown. Slowly bring your fingers together. At what part of the stick do your fingers meet? Can you explain why this always happens, no matter where you start your fingers?

5. Place the hook of a wire coat hanger over your finger. Carefully balance a coin on the straight wire at the bottom directly under the hook. You may have to flatten the wire with a hammer or fashion a tiny platform with tape. With a surprisingly small amount of practice you can swing the hanger and balanced coin back and forth and then in a circle. The tendency of the coin to travel in a straight-line path presses it to the wire. The wire correspondingly presses back, supplying a normal force. We say a *centripetal force* holds the coin in place. Another version of this is nicely done by Joliet physics teacher Bill Blunk. He suspends a

round piece of pizza cardboard by three strings. On the cardboard platform he first places a glass of water and swings it vertically overhead. The water doesn't spill (in fact, carrying beverages in this way is popular in some cultures—a nice way to avoid spillage). Then Bill goes a step further. We've all seen those center of gravity toys where an eagle with outstretched wings (loaded with lead at the tips) balances nicely on the tip of its beak. Place the balanced bird on the platform and it will remain balanced when you swing it overhead. Very impressive.

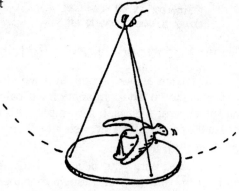

7

Gravity

It would be erroneous to say that Newton discovered gravity. The discovery of gravity goes back much further than Newton's time, to earlier times when earth dwellers found the consequences of tripping and falling. What Newton discovered was that gravity is universal—that it's not a phenomenon unique to earth, as his contemporaries had supposed.

From the time of Aristotle, the circular motion of heavenly bodies was regarded as natural. The ancients believed that the stars, planets, and moon move in divine circles, free from any impelling forces. As far as the ancients were concerned, this circular motion required no explanation. Isaac Newton, however, recognized that a force of some kind must be acting on the planets; otherwise, their paths would be straight lines. Others of his time, influenced by Aristotle, would say any force would be directed along the planet's motion. Newton, however, reasoned that a force on the planets must be perpendicular to their motion—directed toward the center of their curved paths—toward the sun. This, the force of gravity, was the same force that pulls an apple off a tree. Newton's stroke of intuition, that the force between the earth and an apple is the same force that pulls moons and planets and everything else in our universe, was a revolutionary break with the prevailing notion that there were two sets of natural laws: one for earthly events, and another, altogether different for motion in the heavens.

The Universal Law of Gravitation

According to popular legend, Newton was sitting under an apple tree when the idea struck him that gravity extended beyond the earth. Perhaps he looked up through tree branches toward the origin of the falling apple and noticed the moon. In any event, Newton had the insight to see that the force between the earth and a falling apple is the same force that pulls the moon in an orbital path around the earth, a path similar to a planet's path around the sun.

To test this hypothesis, Newton compared the fall of an apple with the "fall" of the moon. He realized that the moon falls in the sense that *it falls away from the straight line it would follow if there were no forces acting on it.* Because of its horizontal speed, it "falls around" the round earth (more about this in the next chapter). By simple geometry the moon's distance of fall per second could be compared to the distance that an apple or anything at that distance would fall in one second. Newton's calculations didn't check. Disappointed, but recognizing that brute fact must always win over a beautiful hypothesis, he placed his papers in a drawer where they remained for nearly 20 years. During this period he founded and developed the field of geometric optics for which he first became famous.

Newton's interest in mechanics was rekindled with the advent of a spectacular comet in 1680 and another two years later. He returned to the moon problem at the prodding of his astronomer friend, Edmund Halley, for whom the second comet was later named. He made corrections in the experimental data used in his earlier method and obtained excellent results. Only then did he publish what is one of the most far-reaching generalizations of the human mind: the **law of universal gravitation.** This is a dramatic example of the painstaking effort and cross-checking that go into the formulation of a scientific theory. Contrast Newton's approach with the failure to "do one's homework," the hasty judgments, and the absence of cross-checking that so often characterize the pronouncements of less-than-scientific theories.

Everything pulls on everything else in a beautifully simple way that involves only mass and distance. According to Newton, every mass attracts every other mass with a force that for any two masses is directly proportional to the product of the masses involved and inversely proportional to the square of the distance separating them.

$$\text{Force} \sim \frac{\text{mass}_1 \times \text{mass}_2}{\text{distance}^2}$$

Expressed symbolically,

$$F \sim \frac{m_1 m_2}{d^2}$$

where m_1 and m_2 are the masses, and d is the distance between their centers. Thus, the greater the masses m_1 and m_2, the greater the force of attraction between them. The greater the distance of separation d, the weaker is the force of attraction—weaker as the inverse square of the distance between their centers of mass.

One of my students found it difficult to accept that gravity gets weaker with distance from the earth, and cited the case that when carrying a heavy load up a flight of stairs, the load got heavier the higher he climbed!

Note the different role of mass in the gravitational equation. Previously we have treated mass as a measure of inertia, which is called *inertial mass*. Now we see mass as a measure of gravitational force, which in this context is called *gravitational mass*. It is experimentally established that the two are equal, and, as a matter of principle, the equivalence of inertial and gravitational mass is the foundation of Einstein's general theory of relativity.

> **Question**
>
> Gravitational force acts on all bodies in proportion to their masses. Why, then, doesn't a heavy body fall faster than a light body?

G: The Universal Gravitational Constant

The proportional form of the universal law of gravitation can be expressed as an exact equation when the constant of proportionality G, called the *universal gravitational constant*, is introduced. Then the equation is

$$F = G\frac{m_1 m_2}{d^2}$$

In words, the force of gravity between two objects is found by multiplying their masses, dividing by the square of the distance between their centers, and then multiplying this result by the constant G. The magnitude of G is the same as the magnitude of the force between two masses of 1 kilogram each, 1 meter apart: 0.0000000000667 newton. This small magnitude of G indicates an extremely weak force. The units of G are such that the force is expressed in newtons. In scientific notation,

$$G = 6.67 \times 10^{-11} \text{ N} \times \text{m}^2 / \text{kg}^2$$

The numerical value of G depends entirely on the units of measurements we choose for mass, distance, and time. The international choices are for mass, the kilogram; for distance, the meter; and for time, the second. G was first measured long after the time of Newton in the eighteenth century by an English physicist, Henry Cavendish. He accomplished this by measuring the tiny force between lead masses with an extremely sensitive torsion balance. A simpler method was later developed by Philipp von Jolly, who attached a spherical flask of mercury to one arm of a sensitive balance. After the balance was put into equilibrium, a 6-ton lead sphere was rolled beneath the

> **Answer**
>
> The reason that a heavy body doesn't fall faster than a light body is because the greater gravitational force on the heavier body (its weight) acts on a correspondingly greater mass (inertia). The ratio of gravitational force to mass is the same for every body—hence all bodies in free fall accelerate equally. (This was illustrated by the falling bricks back in Chapter 3.)

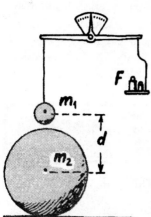

mercury flask. The gravitational force between the two masses was equal to the weight that had to be placed on the opposite end of the balance to restore equilibrium. All the quantities m_1, m_2, F, and d were known, from which the ratio G was calculated:

$$\frac{F}{m_1 m_2 / d^2} = 6.67 \times 10^{-11} \mathrm{N} / \mathrm{kg}^2 / \mathrm{m}^2$$

$$= 6.67 \times 10^{-11} \mathrm{N} \times \mathrm{m}^2 / \mathrm{kg}^2$$

Don't fret over the units here. The magnitude of G tells us that the force of gravity is a very weak force. It is the weakest of the presently known four fundamental forces. (The other three are the electromagnetic force and two kinds of nuclear forces.) We sense gravitation only when masses like that of the earth are involved. The force of attraction between you and a large ship on which you stand is too weak for ordinary measurement. The force of attraction between you and the earth, however, can be measured. It is your weight.

In addition to your mass, your weight also depends on your distance from the center of the earth. At the top of a mountain your mass is no different than it is anywhere else, but your weight is slightly less than at ground level because your distance from the center of the earth is greater.

Once the value of G was known, the mass of the earth was easily calculated. The force that the earth exerts on a mass of 1 kilogram at its surface is 9.8 newtons. The distance between the 1-kilogram mass and the center of mass of the earth is the earth's radius, 6.4×10^6 meters. Therefore, from $F = G(m_1 m_2 / d^2)$, where m_1 is the mass of the earth,

$$9.8\mathrm{N} = 6.67 \times 10^{-11} \mathrm{N} \times \mathrm{m}^2 / \mathrm{kg}^2 \frac{1\,\mathrm{kg} \times m_1}{(6.4 \times 10^6 \mathrm{m})^2}$$

from which the mass of the earth $m_1 = 6 \times 10^{24}$ kilograms.

Physicists like to speculate about the consequences of G having a smaller or a larger value. If G were smaller, gravitation would have been smaller between bits of material in the expanding universe, and material might not have gathered into planets, stars, and galaxies. If G were larger, material would gather more prodigiously, with an altogether different balance between gravitational force that seeks to collapse a body, and electromagnetic force that props it up. Planets might not even exist, for sufficiently stronger gravity would light them up and make them stars. And stars would pull in on themselves more, burn faster, live shorter, with little time for the evolution of life as we know it. How nice that G is just right for the universe we enjoy. We couldn't have it any other way!

Gravity and Distance: The Inverse-Square Law

We can better understand how gravity is diminished with distance by considering how paint from a paint gun spreads with increasing distance. Suppose
we position a paint gun at the center of a sphere with a radius of 1 meter, and
a burst of paint spray travels 1 meter to produce a square patch of paint that is
1 millimeter thick. How thick would the patch be if the experiment were done
in a sphere with twice the radius? If the same amount of paint travels in straight
lines for 2 meters, it will spread to a patch twice as tall and twice as wide. The
paint would be spread over an area four times as big, and its thickness would
be only 1/4 millimeter. Can you see from the figure that for a sphere of radius

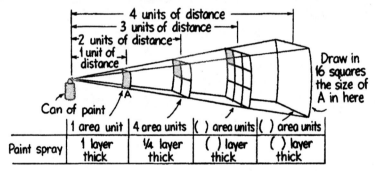

3 meters the thickness of the paint patch would be only 1/9 millimeter? Can you see the thickness of the paint decreases as the square of the distance increases? This is known as the **inverse-square law**. The inverse-square law holds for gravity and for all phenomena wherein the effect from a localized source spreads uniformly throughout the surrounding space: such as the electric field about an isolated electron, light from a match, radiation from a piece of uranium, or sound from a cricket.

In using Newton's equation for gravity, it is important to emphasize that the distance term *d* is the distance between the centers of masses of the objects that are attracted to each other. Note in the figure below that the apple which normally weighs 1 newton at the earth's surface weighs only 1/4 as much when it is twice the distance from the earth's center. The greater the distance from the earth's center, the less the weight of an object. A child that weighs 300 newtons at sea level will weight only 299 newtons atop Mt. Everest. No matter how great the distance, the earth's gravitational force approaches, but never reaches, zero. Even if you were transported to the far reaches of the universe, the gravitational influence of home would still be with you. It may be overwhelmed by the gravitational influences of nearer and/or more massive bodies, but it is there. The gravitational influence of every material object, however small or however far, is exerted through all of space.

This notion that everything pulls on everything is sometimes misapplied, as for example, when the gravitational forces of planets pull on newborn babies. Rather than citing how negligible these forces are in comparison with the effects of gravity of nearer bodies, I have my students calculate the gravitational force between a new-born baby and the planet Mars, and be-

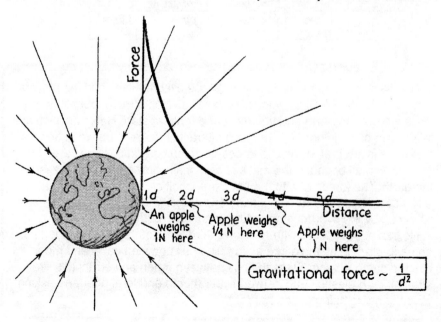

tween the same baby and the doctor that delivers it. Hands down, the gravitational force between the baby and doctor overwhelms the force between the baby and more-distant Mars.

Another misconception about gravity and distance is the notion that the earth's gravitation is zero on objects in space shuttle territory. When I have my students plug into the gravitational equation the additional 200 kilometer altitude of space shuttle territory with the 6380 kilometer radius of the earth, they calculate that the acceleration due to gravity that high is about 94 percent of its surface value. Calculations are often the best way to make a point. To astronauts, although it *seems* as if there's no gravity up there, gravity there is. More about that soon.

How strong is the gravitational force between the earth and the moon? To get a handle on that, I have my students calculate the thickness of a steel cable that would replace gravity if it were shut off. The thickness? About 700 kilometers! Gravity may be weak for small things, but it's quite appreciable between very massive bodies.

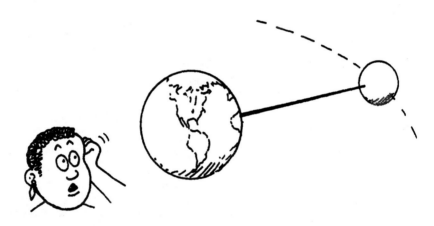

force fields

We know the earth and moon pull on each other. This is *action at a distance,* because the earth and moon interact with each other without being in contact. But we can look at this in a different way: we can regard the moon as in contact with the *gravitational field* of the earth. A gravitational field is the space surrounding a massive body in which another mass experiences a force of attraction. A gravitational field is an example of a **force field,** for any mass in the field space experiences a force. It is common to think of rockets and distant space probes as interacting with the gravitational fields rather than with the masses responsible for these fields. The field concept plays an in-between role in our thinking about the forces between different masses.

Questions

1. By how much does the gravitational force between two objects decrease when the distance between them is doubled? Tripled? Increased tenfold?
2. Light from the sun, like gravity, obeys the inverse square law. If you were on a planet twice as close to the sun, how much brighter would the sun look?
3. Consider an apple at the top of a tree. The apple is pulled by earth gravity with a force of 1 N. If the tree were twice as tall, would the force of gravity be only 1/4 as strong? Defend your answer.

When we study electricity we'll learn about electric fields—the regions surrounding electric charges. We'll also study the *magnetic field* of a magnet (right). Iron filings sprinkled over a sheet of paper on top of the magnet reveal the shape of the magnet's field. The pattern of filings shows the strength and direction of the magnetic field at different points in the space around the magnet. Where the filings are close together, the field is strong. The direction of the filings show the direction of the field at each point. Planet Earth is a giant 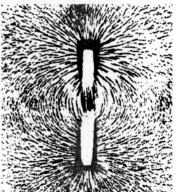 magnet and, like all magnets, is surrounded by a magnetic field. Evidence of the field is easily seen by the orientation of a magnetic compass.

But getting back to gravitational fields, we see the pattern of the earth's  gravitational field can be represented by field lines. Like the iron filings around a magnet, the field lines are closer together where the gravitational field is stronger. At each point on a field line, the direction of the field at that point is along the line. Arrows show the field direction. A particle, astronaut, spaceship, or any mass in the vicinity of the earth will be accelerated in the direction of the field line at that location. The strength of the earth's gravitational field, like the strength of its force on objects, follows

Answers

1. It decreases to one-fourth, one-ninth, and one-hundredth.
2. Four times brighter.
3. No, because the twice-as-tall apple tree is not twice as far from the earth's center. The taller tree would have to have a height equal to the radius of the earth (6,370 km) before the weight of the apple reduces to 1/4 N. Before its weight decreases by 1 percent, an apple or any object must be raised 32 km—nearly four times the height of Mt. Everest. So as a practical matter we disregard the effects of everyday changes in elevation.

the inverse-square law. It is strongest near the earth's surface and weakens with increased distance from the earth.

> **Question**
>
> Strictly speaking, you weigh a tiny bit less when you are in the ground-floor lobby of a massive skyscraper than you do at home. Why?

At the earth's surface the gravitational field varies slightly from location to location. Above large subterranean lead deposits, for example, the field is slightly stronger than average. Above large natural caverns, perhaps filled with natural gas, the field is slightly weaker. Geologists as well as mineral and oil prospectors make precise measurements of the earth's gravitational field to indicate what may lie beneath the surface.

We've said that gravitation is less atop a mountain because of the greater distance to the earth's center. It is lesser still because of the relatively low density of the mountain compared with that of the semi-liquid mantle that the earth's crust floats upon. (Density is mass *per* volume, so equal volumes of continental crust and mantle have different densities—the crust being less

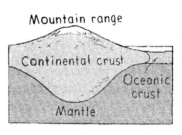

Mountain range

dense.) Less dense things float on more dense fluids. For example, ice has 0.9 the density of water, resulting in only 10% of a floating iceberg above water, with 90% submerged. A mountain similarly floats with only its tip showing. Due to density differences between mountain and mantle, about 85% of a mountain extends below the surrounding ground level. This means the higher-density mantle is pushed far below the surface, far below mountain tops, which further reduces gravitation there.

There is another interesting fact about mountains that merits mentioning: if you could shave off the top of an iceberg, the iceberg would be lighter and be buoyed up to nearly its original height. Similarly, when mountains erode they are lighter and pushed up from below to float to nearly their original heights. So when a kilometer of mountain erodes away, some 85% of it comes back. That's why it takes so long for mountains to wear away.

As previously mentioned, electrical charges are also surrounded by force fields—the electric field, which we shall describe in Book 3. There we'll learn that atoms align with the electric fields of other atoms, and molecules are formed. We'll also learn how magnets align with the magnetic fields of the earth to become compasses. Similar to aligned magnets, we'll soon see how the moon aligns with the earth's gravitational field, which is why only one side of the moon faces us. Force fields have far-reaching effects.

> **Answer**
>
> You weigh a tiny bit less in the lower part of a massive building because the mass of the building above pulls you ever so slightly upward.

Tides

Seafaring people have always known that there is a connection between ocean tides and the moon. Newton was the first to show that tides are caused by *differences* in the gravitational pull between the moon and the earth's opposite sides. Gravitational force between the moon and earth is stronger on the side of the earth nearer to the moon and weaker on the side of the earth farthest from the moon.

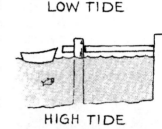

LOW TIDE

HIGH TIDE

To understand why the difference in pulls produces tides, consider a spherical ball of Jell-O. If you exert the same force on every part of the ball, it would remain spherical as it accelerates. But if you pulled harder on one side than the other, there would be a difference in accelerations and the ball would become elongated. That's what's happening to this big ball on which we live. As different pulls of the moon produce a stretching of the earth, the ocean bulges to extend nearly 1 meter above the average surface level of the ocean. The earth rotates once per day, so a fixed point on earth passes beneath both of these bulges each day. This produces two sets of ocean tides per day—two high tides and two low tides. It turns out that while the earth rotates, the moon moves in its orbit and appears at the same position in our sky every twenty-four hours

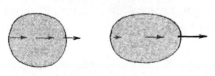

and fifty minutes, so the two-high-tide cycle is actually at twenty-four-hour-and-fifty minute intervals. This means tides do not occur at the same time every day.

The sun also contributes to ocean tides, about half as effective as the moon—even though it pulls 180 times more on the earth than the moon. Why aren't solar tides 180 times greater than lunar tides? Because the *difference* in gravitational pulls on opposite sides of the earth is very small (only about 0.017 percent, compared to 6.7 percent across the earth by the moon).

Proximity is the main factor for tides. Close inspection of the plot of gravity versus distance shown atop page 125 tells it all. Two positions of the earth from the sun are shown. Note that the difference ΔF across the earth is appreciably greater when the earth is near. Likewise between the earth and

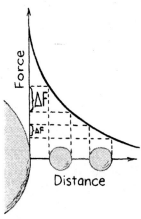

Force

Distance

moon. If the moon were too near the earth, the moon would be pulled apart. This has been the fate of moons too close to other planets—Saturn's rings being the best example.

Newton deduced that the difference in pulls decreases as the *cube* of the distance between the centers of the bodies—twice as far away produces 1/8 the tide; three times as far, only 1/27 the tide, and so on. Only relatively close distances result in appreciable tides, and so our nearby moon out-tides the enormously more massive but farther-away sun. The amount of tide also depends on the size of the body having tides. Although the moon produces a considerable tide in the earth's oceans, which are thousands of kilometers apart, it produces scarcely any in a lake. That's because no part of the lake is significantly closer to the moon than any other part, so there is no significant *difference* in moon pulls on the lake. Similarly for the fluids in your body. Any tides in the fluids of your body caused by the moon are negligible. You're not tall enough for tides. What micro-tides the moon may produce in your body are only about one two-hundredth the tides produced by a one kilogram melon held one meter above your head! Tell this to anyone who claims the gravitational attraction of the moon or that of planets has an influence on humans!

Pseudo-scientists have millions of people believing in the tidal influences of the planets and stars on humans. Rather than preach science to my students, I have them calculate the tidal influences of the moon, the earth, and the one-kilogram melon. They find that the earth produces the most effect, followed by the melon, and then the moon. How much tidal effect on humans does the moon have? About as much as a five-cent coin 1 meter above your head! Still, disturbingly, pseudoscience reigns in popularity. People are fascinated with a make-believe world, when to my mind, the real world is enormously more fascinating. Perhaps junk-science types present their information more interestingly than scientists.

When the sun, earth, and moon are all lined up, the tides due to the sun and the moon coincide, and we have higher-than-average high tides and lower-

than-average low tides. These are called **spring tides**.(Spring tides have nothing to do with the spring season.) Spring tides occur at the times of a new or full moon.

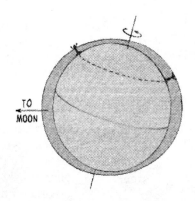

When the moon's phase is half way between a new moon and a full moon, in either direction, the solar and lunar tides partly cancel each other. Then high tides are lower than average and the low tides are not as low as average low tides. These are called **neap tides**.

Because of the earth's tilt, the two tides per day are normally unequal at a given location. The sketch above shows how a person in the Northern Hemisphere may find the tide nearest the moon much lower (or higher) than the tide half a day later. Inequalities of tides vary with the positions of the moon and the sun.

Question

We know that both the moon and the sun produce our ocean tides. And we know the moon plays the greater role because it is closer. Does its closeness mean it pulls with more gravitational force than the sun on the earth's oceans?

Because much of the earth is molten we have earth tides, though less pronounced than ocean tides. There are also atmospheric tides, which regulate the cosmic rays that reach the earth's surface. Our brief treatment of tides is quite simplified, for the tilt of the earth's axis, interfering land masses, friction with the ocean bottom, and other factors complicate tidal motions. Tides are fascinating!

Answer

No, the sun's pull is much stronger. Gravitational pull weakens as the distance squared to the body that pulls. But the *difference* in pulls across the earth's oceans weakens as the distance cubed. When the distance to the sun is squared, gravitation from the sun is still stronger than gravitation from the closer moon—because of the sun's enormous mass. But when the distance to the sun is cubed, as is the case for tidal forces, the sun's influence is less than the moon's. Distance is the key to tidal forces. If the moon were closer to earth, the tides on both the earth and the moon would increase as the closer distance cubed, which could tear the moon into pieces—catastrophic—as the planetary rings of other planets suggest.

Tides on the Moon

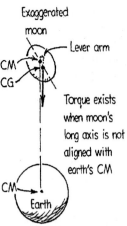

Exaggerated moon

There are two tidal bulges on the moon for the same reason there are two tidal bulges on the earth—near and far sides of each body are pulled differentially. So the moon is pulled slightly out of a spherical shape into a football shape, with its long axis pointing toward the earth. But unlike earth's tides, the tidal bulges are stationary, with no "daily" rising and falling of moon tides. Since the moon takes 27.3 days to make a single revolution about its own axis (and also about the earth-moon axis), the same part of its surface faces the earth all the time. This is because the elongated moon's center of gravity is slightly displaced from its center of mass, so whenever the moon's long axis is not pointed toward earth, the earth exerts a small torque on the moon. This tends to twist the moon toward aligning with the earth's gravitational field, like the torque that aligns a compass needle with a magnetic field. So we see there is a reason why the moon always shows us its same face!

Remarkably, this "tidal lock" is also working on the earth. Our days are getting longer at the rate of 2 milliseconds per century. In a few billion years our day will be as long as a month and the earth will always show the same face to the moon. How about that?

Questions
1. If somebody tugged on your shirt sleeve, it stands a good chance of tearing. But if all parts of your shirt are tugged equally, no tearing would occur. How does this relate to tidal forces?
2. If the moon didn't exist, would the earth still have ocean tides? If so, how often?
3. What would be the effect on the earth's tides if the diameter of the earth were very much larger than it is? If the earth were as it presently is, but the moon very much larger than it is with its same mass?

Answers
1. Just as differences in tugs on your shirt will distort the shirt, differences in tugs on the oceans distort the ocean and produce tides.
2. Yes, the earth's tides would be due only to the sun. They'd occur twice per day (every 12 hours instead of every 12.5 hours) due to the earth's daily rotation.
3. Tides would be greater if the earth's diameter were greater because the difference in pulls would be greater. Tides on earth would be no different if the moon's diameter were larger. The gravitational influence of the moon is as if all the moon's mass were at its center, or CG. Tidal bulges on the solid surface of the moon, however, would be greater if the moon's diameter were larger—but not on the earth.

Weight and Weightlessness

When you step on a spring balance like a bathroom scale, you compress a spring inside. When the pointer stops, the strong electrical forces between the molecules inside the spring material balance the gravitational attraction between you and the earth—nothing moves. You and the scale are in static equilibrium. The pointer is calibrated to show your weight. If you stood on a bathroom scale in an accelerating elevator, you'd find variations in your weight. If the elevator accelerated upward, the springs inside the bathroom scale would be more compressed and your weight reading would increase. If the elevator accelerated downward, the springs inside the scale would be less compressed and your weight reading would decrease. If the elevator cable broke and the elevator fell freely, the reading on the scale would go to zero. According to the reading, you would be weightless. Would you really be weightless? We can answer this question only if we agree on what we mean by *weight*.

Recall in Chapter Three we defined **weight** as the gravitational force exerted on an object by the nearest most massive body. According to this definition, you would have weight whether or not you were falling, for you are still gravitationally attracted to the earth. So your weight and the weight you experience, your *apparent weight*, can be very different. We define **apparent weight** as the force an object exerts against the supporting floor or the weighing scales. According to this definition, you are as heavy as you feel; so in an elevator that accelerates downward, the supporting force of the floor is less and your apparent weight is less. If the elevator is in free fall, your apparent weight is zero. Even in this weightless condition, however, there is still a gravitational force acting on you, causing your downward acceleration. But gravity now is not felt as weight because there is no support force in the falling elevator.

Suppose you're an astronaut in orbit. You feel weightless because you are not supported by anything. There would be no compression in the springs of a bathroom scale placed beneath your feet because the bathroom scale is falling as fast as you are. (We'll see in the next chapter that a satellite is an object in free fall; it's sideways motion is great enough to insure that it falls around the earth rather than into it.) If you drop an object in your vicinity, say a pocket pen, you drop with the pen and it remains in your vicinity. Local effects of gravity seem to be eliminated. Your body organs respond as though gravitational forces were absent, and this gives the sensation of weight-lessness. You experience the same sensation in orbit that you'd feel in a falling elevator or in a car driving off a cliff—a state of free fall.

Several years ago when flying back from an AAPT meeting where astronaut Sally Ride was a guest speaker, I had the good fortune of sitting one seat away from her. I was fascinated by her discussion of what it felt like to be in orbit, and the standard fare of barf bags for beginning astronauts. In orbit, you're really in a state of continual free fall, which takes some time to get used to.

All the while you're in orbit with zero apparent weight, you're still under the influence of gravity. To be truly weightless, you'd have to be far enough out in space where gravitational forces would be negligible—well away from the earth, sun, and other attracting bodies. In this truly weightless environment, any motion would be in a straight-line path rather than the curved path of a closed orbit.

Question

Pretend you're in an elevator at the top of a tall building. Mounted in the elevator is a video camera that takes pictures of you holding a pen in front of your face, then dropping the pen. If this is done at the same time that the elevator cable snaps, so the elevator freely falls, how will the video footage of you dropping the pen be similar to footage of you dropping the same pen while in the orbiting space shuttle?

Black Holes

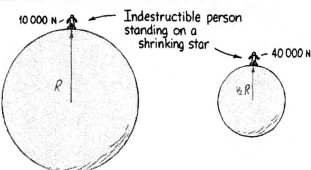

Suppose you were indestructible and could travel in a spaceship to the surface of a star. Your weight would depend both on your mass and the star's mass and on the distance between the star's center and your belly button. If the star were to burn out and collapse to half size with no change in its mass, your weight at its surface, determined by the inverse-square law, would be four times as much. If the star collapsed to a tenth its size, your weight at its surface would be 100 times as much. If the star kept shrinking, the gravitational field at the surface would become stronger. It would be more and more difficult for a spaceship to leave. The velocity required to escape, *escape velocity*, would increase. If a star such as our sun collapsed to a radius of less than 3 kilometers, the escape velocity from its surface would exceed the speed of light, and nothing—not even light—could escape! The sun would be invisible. It would be, as physicist John Archibald Wheeler named it, a *black hole*. It's not a hole like an opening, but a hole in the sense that matter can fall into it. Gravitation near these shrunken stars is so enormous that light cannot escape from their vicinity. Black holes eat their own light!

Although we'd get no light from the sun if it collapsed to become a black hole, its gravitational field here at the earth's distance, quite amazingly, would be unchanged! Why? Because nothing in the equation for gravity changes. Masses are the same (although the sun's would be quite compressed), and distance from center to center would remain the same. Interestingly, the gravitational fields of black holes are the same as those of the stars before collapse—at distances beyond the original star radii. So, contrary to widespread belief, black holes are non-aggressive and don't reach out and swallow innocents at a distance. Only at distances closer than the initial stellar radius are the gravitational fields more enormous than before collapse.

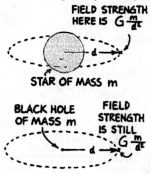

FIELD STRENGTH HERE IS $G\frac{m}{d^2}$

STAR OF MASS m

BLACK HOLE OF MASS m   FIELD STRENGTH IS STILL $G\frac{m}{d^2}$

The sun, in fact, probably has too little mass to experience such a collapse, but when some stars with greater mass—now estimated to be at least 1.5 solar masses or more—reach the end of their nuclear resources, they undergo collapse; and unless rotation is fast enough, the collapse continues until the stars reach infinite densities. These shrunken stars have crushed themselves out of visible existence—they're completely invisible. In Einstein's model for gravity, a black hole is a warp in the fabric of space itself, much like the surface of a waterbed is warped when a heavy weight is put on it.

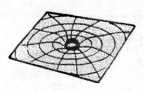

Answer

In both cases, the pen would be seen hovering in front of your face. To many people, the hovering pen is seen as evidence of no gravity. Perhaps seeing similar scenes broadcast from the orbiting space shuttles is a strong reason for many people believing there is no gravity in outer space.

Far from a black hole, the warp of space is slight. But near a black hole, at distances closer than the star's original radius, the collapse of space is significant. The surrounding warp draws anything that passes too close— light, dust, or a spaceship. Astronauts could enter the fringes of this warp and with a powerful spaceship still escape. After a certain distance, however, they could not escape, and they'd disappear from the observable universe. Any object falling into a

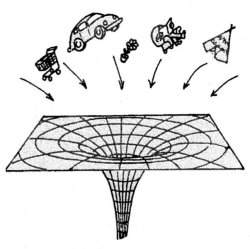

black hole would be torn to pieces. No feature of the object would survive except its mass, its angular momentum (if any), and its electric charge (if any).

How can a black hole be detected, if there is literally no way to "see" it? It makes itself felt by its gravitational influence upon neighboring stars. There is now good evidence that some binary star systems consist of a luminous star and an invisible companion with black-hole-like properties orbiting around each other. Even stronger evidence for more massive black holes is observed in the central regions of some other galaxies. There, stars are circling in a powerful gravitational field around an apparently empty center. These black holes may have masses more than one billion times the mass of our sun. Such a black hole likely occupies the center of our own galaxy. Discoveries are coming faster than books can report. Check with astronomy buffs or a web site for the latest update.

A theoretical entity with some similarity to a black hole is the "wormhole." Like a black hole, a wormhole is an enormous distortion of space and time. But instead of collapsing toward an infinitely dense point, the wormhole opens out again in some other part of the universe—or even, conceiv-

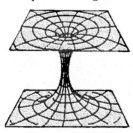

ably, in some *other* universe! Whereas the existence of black holes has been confirmed through experiment, the wormhole remains speculative. But since the same laws of physics that explain the black hole also predict the possibility of wormholes, it won't be surprising if someday their existence is confirmed. Some physicists imagine that wormholes open the possibility of time travel.

Question

An unfortunate astronaut falling into a black hole would probably be killed by tidal forces before ever encountering the hole itself. Why?

Universal Gravitation

We all know that the earth is round. But why is it round? It is round because of gravitation. Everything attracts everything else, and so the earth has attracted itself together as much as it can! Any "corners" of the earth have been pulled in; as a result, every part of the surface is equidistant from the center of gravity. The tightly pulled surface has minimum area for the volume contained. The shape is a sphere. Therefore, we see from the law of gravitation that the sun, the moon, and earth are spherical because they have to be (rotational effects make them slightly ellipsoidal).

If everything pulls on everything else, then the planets must pull on each other. The force that acts on Jupiter, for example, is not just the force from the sun; there are also pulls from the other planets. Their effect is small in comparison to the pull of the much more massive sun, but it still shows. When Saturn is near Jupiter, its pull disturbs the otherwise smooth ellipse traced by Jupiter. Both planets "wobble" about their expected orbits. This wobbling is called a *perturbation*. By the 1840s, studies of the most recently discovered planet, Uranus, showed that the deviations of its orbit from a perfect ellipse could not be explained by perturbations from all other known planets. Either the law of gravitation was failing at this great distance from the sun or an unknown eighth planet was perturbing Uranus. An Englishman, J. C. Adams, and a Frenchman, Urbain Leverrier, each assumed Newton's law to be valid and independently calculated where an eighth planet should be. At about the same time, both sent letters to their respective observatories with instructions to search a certain area of the sky. The request by Adams was delayed by misunderstandings at Greenwich, but Leverrier's request to the director of the Berlin observatory was heeded immediately. The planet Neptune was discovered that very night!

Other perturbations of the planet Uranus led to the prediction and discovery of the ninth planet, Pluto, in 1930 at the Lowell Observatory in Arizona. Pluto takes 248 years to make a single revolution about the sun, so no one will see it in its discovered position again until the year 2178.

The shapes of distant galaxies provide further evidence that the law of gravitation applies to larger distances—even underlying the fate of the entire universe. Current scientific speculation is that the universe originated in the explosion of a primordial fireball some 8 to 15 billion years ago. This is the *Big Bang* theory of the origin of the universe. The explosion was space itself, with all the matter of the universe hurled outward. Early waves of light have been stretched out to fill the present universe with a cosmic microwave background. Space is

---

Answer

The part of the person nearest the hole would be pulled appreciably harder than parts farther from the black hole. This would produce an enormous tidal force across the person's body.

still stretching out, carrying the galaxies with it. This expansion may go on indefinitely, or it may eventually be overcome by the combined gravitation of all the galaxies and come to a halt. Like a stone thrown upward, whose departure from the ground comes to an end when it reaches the top of its trajectory and which then begins its descent to the place of its origin, the universe may contract and fall back into a single unity. This would be the *Big Crunch*. After that, we can only speculate that the universe might re-explode to produce a new universe. The same course of action might repeat itself, and the process may well be cyclic. If this speculation is true, we live in an oscillating universe.

If the universe does oscillate, who can say how many times this process has repeated? We know of no way a civilization could leave a trace of ever having existed, for all the matter in the universe would be reduced to bare subatomic particles or new entities during such an event. We can only speculate that formation of the elements, stars, galaxies, and life again takes place. All the laws of nature, such as the law of gravitation, might then be rediscovered by the higher-evolving life forms. Then students of these laws might read about them, as you are doing now. Think about that!

We do not know whether the expansion is indefinite because we are uncertain about whether enough mass exists to halt the expansion. If the expansion halts and is followed by contraction, the time from Big Bang to Big Crunch is estimated to be somewhat less than 100 billion years. Now many cosmologists think the expansion is picking up speed with something other than gravity acting over large distances. In this scenario the universe will expand forever. We don't know. Our universe is still young. But humankind is far, far younger.

Quite amazingly, there's a lot more mass out there than we can directly see. Astrophysicists talk of the **dark matter**—unseen matter that tugs on stars and galaxies that we *can* see. Gravitational forces within galaxies are measured to be far greater than visible matter can account for. Dark matter is estimated to make up some 90 percent of the mass of the universe. Whatever it is, some, most, or all of it is likely to be "exotic" matter—very different from the elements that make up the periodic table of the elements. It seems to be different stuff. At this writing, the dark matter is not yet identified. Speculations abound, but we don't know what it is.

Few theories have affected science and civilization as much as Newton's theory of gravitation. The successes of Newton's ideas ushered in the so-called Age of Reason (also called the Century of Enlightenment), for Newton had demonstrated that by observation and reason, by employing mechanical models, and by deducing mathematical laws, people could uncover the very workings of the physical universe. How profound that all the moons and planets and stars and galaxies have such a beautifully simple rule to govern them:

$$F = G\frac{m_1 m_2}{d^2}$$

The formulation of this simple rule is one of the major reasons for successes in sciences that followed, because it provided hope that other phenomena of the world might also be described by equally simple laws.

This hope nurtured the thinking of many scientists, artists, writers, and philosophers of the 1700s. One of these was the English philosopher John Locke, who argued that observation and reason, as demonstrated by Newton, should be our best judge and guide in all things, and that all of nature and even society should be searched to discover any natural laws that might exist. Using Newtonian physics as a model of reason, Locke and his followers modeled a system of government that found adherents in the thirteen British colonies across the Atlantic. These ideas culminated in the Declaration of Independence and the Constitution of the United States of America.

Try It and See

Hold your hands outstretched, one twice as far from your eyes as the other, and make a casual judgment as to which hand looks bigger. Most people see them to be about the same size, while many see the nearer hand as slightly bigger. Almost nobody upon casual inspection sees the nearer hand as four times as big. But by the inverse-square law, the nearer hand should appear twice as tall and twice as wide and, therefore, occupy four times as much of your visual field as the farther hand. Your belief that your hands are the same size is so strong that you likely overrule this information. Now if you overlap your hands slightly and view them with one eye closed, you'll see the nearer hand as clearly bigger. This raises an interesting question: what other illusions do you have that are not so easily checked?

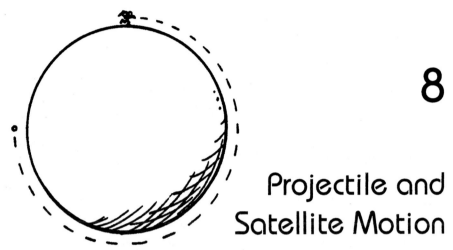

8

Projectile and Satellite Motion

From the top of Mauna Kea in Hawaii (or any high vantage point where the distant horizon is sharp and clear) you can see the curvature of the earth. You have to eyeball the line where ocean and sky meet against a long straightedge in front of your eyes. Otherwise you can't be sure if your eyes are playing tricks on you. Line up your sight so that the bottom edge of the middle of the straightedge just touches the juncture between sky and ocean, and you'll note a space between sky and ocean at the ends. You're seeing the earth's curvature. Now toss a rock horizontally toward the horizon. It quickly falls several meters to the ground below in front of you. It curves as it falls. You'll note that the faster you throw the rock, the wider the curve. Then you wonder how fast Superman would have to throw the rock to clear the horizon ahead. And how fast he'd have to throw it so that its curve matched the curve of the earth. For if he could do that, and air drag were somehow eliminated, the rock would follow a curved path completely around the earth and become an earth satellite! A satellite is, after all, no more than a projectile moving fast enough to continually clear the horizon as it falls.

Projectile Motion

Without gravity, you could toss a rock at an angle skyward and it would follow a straight-line path. Because of gravity, however, the path curves. A tossed rock, a cannonball, or any object that is projected by some means and continues in motion by its own inertia is called a **projectile**. To the early cannoneers, the curved paths of projectiles seemed hopelessly complex. Now we see these

135

paths are surprisingly simple when we look at the horizontal and vertical components of motion separately.

The horizontal component of motion for a projectile is no more complex than the horizontal motion of a bowling ball rolling freely along a level bowling alley. If the retarding effect of friction can be ignored, the bowling ball moves at constant velocity. Since there is no force acting horizontally on the ball, it rolls of its own inertia and covers equal distances in equal intervals of time. It rolls without accelerating. The horizontal component of a projectile's motion is just like the bowling ball's motion along the alley.

The vertical component of motion for a projectile following a curved path is just like the motion described in Chapter 2 for a freely falling object. Like a ball dropped in mid-air, the projectile moves in the direction of earth gravity and accelerates downward. The faster the object falls, the greater the distance covered in each successive second. Or if projected upward, the vertical distances of travel become less with time on the way up.

The curved path of a projectile is a combination of horizontal and vertical motion. The horizontal component of motion for a projectile is completely independent of the vertical component of motion. So unless air drag or some other horizontal force acts, the constant horizontal velocity component is not affected by the vertical force of gravity. Each component acts independently of the other. Their combined effects produce the curved paths of projectiles.

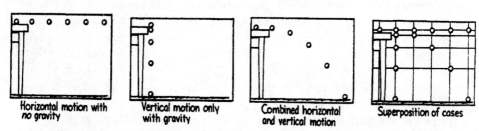

Horizontal motion with no gravity   Vertical motion only with gravity   Combined horizontal and vertical motion   Superposition of cases

These ideas are neatly illustrated in the simulated multiple-flash exposure shown in the figure above, which shows equally-timed sequential positions for a ball rolling off the edge of a level table. Investigate it carefully, for there's a lot of good physics there. The curved path of the ball is best analyzed by considering the horizontal and vertical components of motion separately. There are two important things to notice. The first is that the ball's horizontal component of motion doesn't change as the falling ball moves forward. The ball travels the same horizontal distance in equal times between each flash. That's because there is no component of gravitational force acting horizontally. Gravity acts only *downward*, so the only acceleration of the ball is *down-*

ward. The second thing to note is that the vertical positions become farther apart with time. The vertical distances traveled are the same as if the ball were simply dropped. Note the curvature of the ball's path is the combination of horizontal motion that remains constant, and vertical motion that undergoes acceleration due to gravity.

> **Question**
>
> At the instant a horizontally held rifle is fired over level ground, a bullet held at the side of the rifle is released and drops to the ground. Ignoring air resistance, which bullet, the one fired downrange, or the one dropped from rest, strikes the ground first?

Consider a cannonball shot at an upward angle. Pretend for a moment that there is no gravity; according to the law of inertia, the cannonball would follow the straight-line path shown by the dashed line. But there *is* gravity, so this doesn't happen. What really happens is that the cannonball continually falls beneath the imaginary line until it finally strikes the ground. Get this: like the ball that rolls off the table, the vertical distance it falls beneath any point on the dashed line is the same vertical distance it would fall if it were dropped from rest at that position, and had been falling for the same amount of time. This distance, as introduced in Chapter Two, is given by $d = 1/2\, gt^2$, where t is the elapsed time.

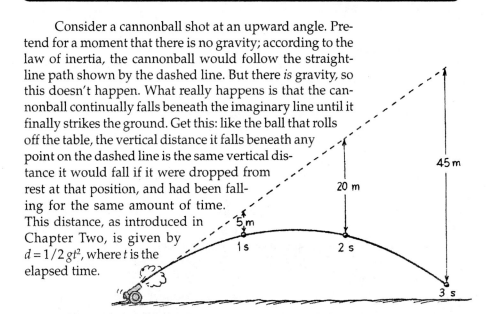

> **Answer**
>
> Both bullets fall the same vertical distance with the same acceleration g due to gravity and therefore strike the ground at the same time. Can you see that this is consistent with our analysis of the figures above? We can reason this another way by asking which bullet would strike the ground first if the rifle were pointed at an upward angle. In this case, the bullet that is dropped would hit the ground first. Now consider the case where the rifle is pointed downward. The fired bullet hits first. So upward, the dropped bullet hits first; downward, the fired bullet hits first. There must be some angle at which there is a dead heat—where both hit at the same time. Can you see it would be when the rifle is neither pointing upward or downward—when it is horizontal?

We can express this another way: shoot a projectile skyward at some angle and pretend there is no gravity. After so many seconds t, it should be at a certain point along a straight-line path. But because of gravity, it isn't. Where is it? The answer is that it's directly below this point. How far below? The answer in meters is $5t^2$. Yum!

> **Questions**
>
> 1. Suppose the cannonball was fired faster. How many meters below the dashed line would it be at the end of the 5 s?
>
> 2. If the horizontal component of the cannonball's velocity were 20 m/s, how far downrange would the cannonball be at the end of 5 s?

The question about the times of fall for the fired and dropped bullets is a favorite in college physics classes. My friend Dave Vasquez asks the same question of middle school kids and guides them to full understanding of the independence of vertical and horizontal motions. He does this with CD software he has developed in which he makes a game of learning physics. One game features mountain climbers needing supplies while stranded on a remote peak. The game player has to compute correct parameters that will drop the supplies in the correct place from an airplane flying above. The plane has only one package, and the dropping area is small. Dave finds that the young players take on the challenge, make the necessary computations, and have a high success rate—one high enough to please most college physics professors.

The sketch of the girl tossing a ball shows the velocity of a projectile resolved into horizontal and vertical components. Only the vertical component changes with time. When rising, it decreases with time as it goes *against* gravity. When descending, it increases with time as it travels *with* gravity. Only the horizontal component is constant—for it neither goes against nor with gravity. In the absence of air resistance, there is no horizontal force, nor a

> **Answers**
>
> 1. Assuming $g = 10$ m/s², the vertical distance beneath the dashed line at the end of 5 s would be 125 m [$d = 5t^2 = 5(5)^2 = 5(25) = 125$]. Interestingly enough, this distance doesn't depend on the angle of the cannon. If air resistance is neglected, any projectile will fall a vertical distance $5t^2$ meters below where it would have reached if there were no gravity.
>
> 2. In the absence of air resistance, the cannonball will travel a horizontal distance of 100 m [$d = vt = (20)(5) = 100$]. Note that since gravity acts only vertically and there is no acceleration in the horizontal direction, the cannonball travels equal horizontal distances in equal times. This distance is simply its horizontal component of velocity multiplied by the time (and not $5t^2$, which applies only to vertical motion under the acceleration of gravity).

horizontal compo-
nent of gravity to
change it.

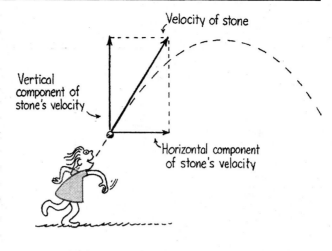

Note the hori-
zontal and vertical
components of ve-
locity for the projec-
tile at different
points in its trajec-
tory (below). We see
the horizontal com-
ponent is the same
everywhere, and
only the vertical
component changes.
Note also that the actual velocity is represented by the vector that forms the
diagonal of the rectangle formed by the vector components. At the top of the
trajectory the vertical component vanishes to zero, so the actual velocity there
is solely the horizontal component of velocity. Everywhere else the magnitude of velocity is greater (just as the diagonal of a rectangle is greater than
either of its sides).

On the next page, we see the paths of several projectiles in the absence
of air resistance, all with the same initial speed but different projection angles.
Notice that these projectiles reach different *altitudes*, or heights above the
ground. They also have different *ranges*, or distances traveled horizontally.
The remarkable thing to note is that the same range is obtained from two
different projection angles—angles that add up to 90 degrees! An object thrown
into the air at an angle of 60 degrees, for example, will have the same range as
if it were thrown at the same speed at an angle of 30 degrees. For the smaller
angle, of course, the object remains in the air for a shorter time.

We have emphasized projectile motion without air resistance. You can
neglect air resistance for a ball you toss back and forth with your friends because the speed is small. Speed makes a difference. Air resistance is a factor
for high-speed projectiles and both range and altitude are less.

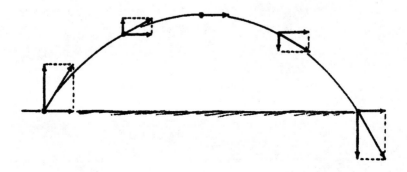

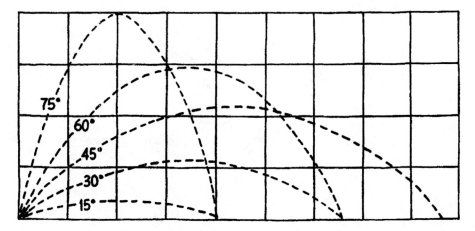

When air resistance is low enough to be negligible, a projectile will rise to its maximum height in the same time it takes to fall from that height to the ground. This is because its deceleration by gravity while going up is the same as its acceleration by gravity while coming down. The speed it loses while going up is, therefore, the same as the speed it gains while coming down. So the projectile arrives at the ground with the same speed it had when it was projected from the ground. Howie Brand, dear friend since my college days,

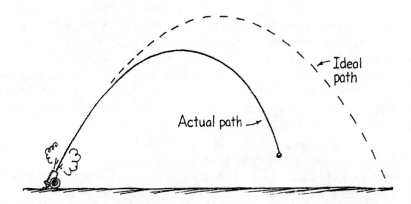

Questions

1. A projectile is launched at an angle into the air. If air resistance is negligible, what is the acceleration of its vertical component of motion? Of its horizontal component of motion?

2. At what part of its trajectory does a projectile have minimum speed?

3. A ball tossed into the air undergoes acceleration while it follows a parabolic path. When the sun is directly overhead, does the shadow of the ball across the field also accelerate?

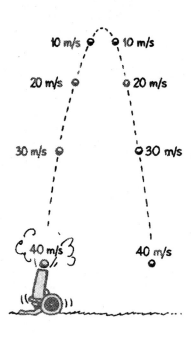

likes to ask his high-school students about the force that acts on the projectile during flight. He finds they don't always connect the constant acceleration with the same constant force—that due to gravity. It takes time to make all these connections.

An athlete or dancer jumping upward is a projectile as soon as the feet leave the ground. How far the jumper goes horizontally has to do with horizontal speed when the jump is executed plus hang time. Recall from Chapter 2 that hang time depends only on the vertical speed when the feet leave the ground. In practice, one *can* increase the vertical component in a running jump and, therefore, *can* jump higher. But this is only because the vertical component of lift-off is increased after the foot bounds against the ground during lift-off. Once lift-off is achieved, horizontal and vertical components of projectile motion are independent of each other.

Baseball games normally take place on level ground. For the short-range projectile motion on the playing field, the earth can be considered to be flat because the flight of the baseball is not affected by the earth's curvature. For very long range projectiles, however, the curvature of the earth's surface must be taken into account. When an object is projected fast enough, it will fall all the way around the earth and become a satellite.

Answers

1. The acceleration in the vertical direction is g because the force of gravity is along the vertical direction. (Recall from Chapter 3 that acceleration is always in the direction of the force that acts on an object.) The acceleration is zero in the horizontal direction because no horizontal force acts on the projectile.

2. The speed of a projectile is minimum at the top of its path. If it is launched vertically, its speed at the top is zero. If it is projected at an angle, the vertical component of speed is zero at the top, leaving only the horizontal component. So the speed at the top is equal to the horizontal component of the projectile's velocity at any point.

3. No, for the shadow moves at constant velocity across the field, showing exactly the motion due to the horizontal component of the ball's velocity.

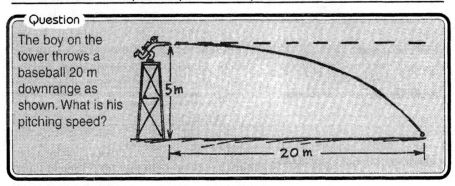

Question

The boy on the tower throws a baseball 20 m downrange as shown. What is his pitching speed?

5 m

20 m

Satellite Motion

Consider the baseball pitcher in the previous check question. If gravity did not act on the ball, the ball would follow a straight line path shown by the dashed line. But gravity does act, so the ball falls below this straight line path. In fact, 1 second after the ball leaves the pitcher's hand, it will have fallen a vertical distance of 5 meters below the dashed line—whatever the pitching speed. It is important to understand this, for it is the crux of satellite motion.

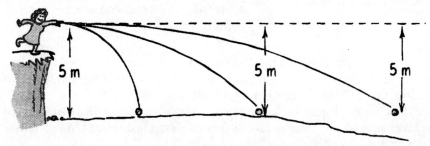

5 m   5 m   5 m

As previously stated: a satellite of the earth is simply a projectile that falls *around* the earth rather than *into* the earth. The speed of the satellite must be great enough to ensure that its falling distance matches the earth's curvature.* It is a geometrical fact about the earth's curvature that its surface

Answer

The ball is thrown horizontally, so the pitching speed is the horizontal distance divided by time. A horizontal distance of 20 m is given, but the time is not stated. However, while the ball is moving horizontally at constant speed, it falls due to gravity a vertical distance of 5 m, which takes 1s. So pitching speed $v = d/t = (20m)/(1s) = 20$ m/s. It is interesting to note that consideration of the equation for constant speed, $v = d/t$ guides thinking about the crucial factor in this problem—the time.

---

* The conventional definition of *to fall* is "to get closer to the earth"; satellites such as the moon do not do this. In science we will find many cases where the technical definition differs from the conventional. For example, we say "the sun sets" and "the moon rises," but technically they do not.

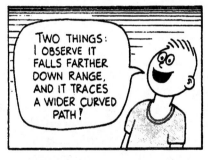

drops a vertical distance of 5 meters for every 8000 meters tangent to the surface (a *tangent* to a circle or to the earth's surface is a straight line that touches the circle or surface at only one place, so is parallel to the circle at the point of contact). This means that if you were floating in a calm ocean, you would be able to see only the top of a 5-meter mast on a ship 8 kilometers away. So if a

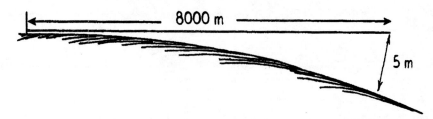

baseball could be thrown fast enough to travel a horizontal distance of 8 kilometers during the time (1 second) it takes to fall 5 meters, then it would follow the curvature of the earth (neglecting air drag). A little thought will show that this speed is 8 kilometers per second. If this doesn't seem fast, convert it to kilometers per hour and you get an impressive 29,000 kilometers per hour (or 18,000 miles per hour)!

At this speed atmospheric friction would incinerate the baseball, or even a chunk of iron. This is the fate of grains of sand and other meteorites that graze the earth's atmosphere, burn up, and appear as "falling stars." That is why satellites like the space shuttles are launched to altitudes higher than 150 kilometers, above the atmosphere. It is a common misconception that satellites orbiting at high altitudes are free from gravity. Nothing could be further from the truth. The force of gravity on a satellite 150 kilometers above the earth's surface is nearly as great as at the surface. High altitude puts the satellite beyond the earth's atmosphere, but not beyond the earth's gravity.

Satellite motion was understood by Newton, who reasoned that the moon was simply a projectile circling the earth under the attraction of gravity. This concept is illustrated in a drawing by Newton, shown below. He compared motion of the moon to a cannonball fired from the top of a high mountain. He imagined that the mountain top was above the earth's atmosphere, so that air resistance would not impede the motion of the cannonball. If a cannonball were fired with a low horizontal speed, it would follow a curved path and soon hit the earth below. If it were fired faster, its path would be less curved and it would hit the earth farther away. If the cannonball were fired fast enough, Newton reasoned, the curved path would become a circle and the cannonball would circle the earth indefinitely. It would be in orbit.

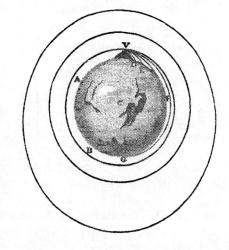

Both cannonball and moon have "sideways" velocity, (*tangential velocity*)—the velocity parallel to the earth's surface—sufficient to ensure motion *around* the earth rather than *into* it. If there is no resistance to reduce its speed, the moon or any earth satellite "falls" around and around the earth indefinitely. Similarly with the planets that continually fall around the sun in closed paths. Why don't the planets crash into the sun? They don't because of their tangential velocities. What would happen if their tangential velocities were reduced to zero? The answer is simple enough: their motion would be straight toward the sun and they would indeed crash into it. Any objects in the solar system without sufficient tangential velocities have long ago crashed into the sun. What remains is the harmony we observe.

Calculating Satellite Speed

For those lecturing on satellite motion, I highly recommend the following skit that guides students into calculating the speed necessary

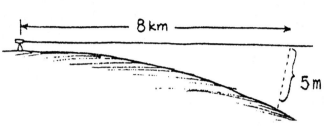

for close earth orbit: consider a horizontal laser standing about a meter above the ground with its beam shining across a level sandy desert. The beam is straight, but the desert floor curves 5 m over an 8000-m or 8-km tangent, which you sketch on the chalkboard (or overhead projector) as above. This is nearly the illustration of the earth's curvature of the previous page. Of course neither are to scale.

Now erase the laser, and sketch in a super cannon positioned so that it points along the laser line. We call this Newton's cannon, for he first suggested its use for describing satellite motion. Consider a cannonball fired at say, 2 km/s, and ask how far downrange will it be at the end of one second, ignoring air drag. Class response should yield an answer of 2 km, which you indicate as below. But it doesn't really get to this place, you say, for it falls beneath the straight line because of gravity. How far? 5 m if the sandy ground weren't in the way. Ask if 2 km/s is sufficient for orbiting the earth. Clearly

not, for the cannonball strikes the sandy ground. If the cannonball is not to hit the ground, we'd have to dig a trench first, as you show on your sketch, which now looks like this:

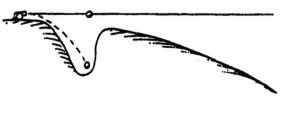

Continue by considering a greater muzzle velocity, say 4 km/s, so the cannonball travels 4 km in one second. Ask if this is fast enough to attain an earth orbit. Student response should indicate that they realize that the cannonball will hit the ground before one second is up. Then repeat the previous line of reasoning, again having to dig a trench, and your sketch should look like this:

Continue by considering a greater muzzle velocity—great enough so that the cannonball travels 6 km in one second. This is 6 km/s. Ask if this is fast enough not to hit the ground (or equivalently, if it is fast enough for earth orbit!) Then repeat the previous line of reasoning, showing that the cannonball "eats sand" unless you've dug a trench. Now your sketch looks like this:

You're almost there! Continue by considering a muzzle velocity great enough so that the cannonball travels 8 km in one second. (Don't state the velocity is 8 km/s here as you'll diminish your punch line.) Repeat your previous reasoning and note that this time you don't have to dig a trench! After a pause, you ask what the speed must be to not "eat sand" and to orbit the earth. Done properly, you have led your class into a "derivation" of orbital speed about the earth with no equations or algebra.

I sum this up by asking students to imagine they were fired horizontally at 8 kilometers per second from a cannon several meters above the ground. In the one second it would take for them to travel 8 kilometers, they'd still be the same several meters above the ground, for they'd fall a vertical distance of 5 meters while the earth would have curved downward 5 meters beneath them. After another second, if they continued at the same speed and the ground remained perfectly flat, they'd be another 8 kilometers down range, but still the same height above the ground. If there were no air drag they'd maintain their speed, and, if they didn't run into any obstacles, they'd follow the curve of the earth, falling continually while remaining at a constant altitude—they'd be in low-earth orbit!

Gravitational force is less on satellites in higher orbits so they do not need to go so fast. (Speed for circular orbit is $v = \sqrt{GM/d}$, so a satellite at four times the earth's radius needs to travel only half as fast, 4 km/s.) The moon is very far away and travels considerably slower, which is why it takes a month to circle the earth.

Everybody loves to see the demonstration of a bucket of water swung in a vertical circle. The explanation for why the water doesn't spill is analogous to

the explanation for why satellites don't "spill" to earth. This demo usually illustrates centripetal or centrifugal force (not covered in this book). The most straightforward explanation as to why the water doesn't spill doesn't involve these terms. It so happens that the water at the top of the swing *does* fall, and the execution of the swing is simply to make the bucket fall at least as fast as the water. This is why the water doesn't spill. In a similar way, the orbiting space shuttle *does* fall. The distance it falls is matched by the distance the earth curves, so the satellite remains above its surface. Analogies are the way to understand concepts!

Circular Orbits

So the tangential velocity needed for close-earth orbit is 8 kilometers per second. An 8-kilometers-per-second cannonball fired horizontally from Newton's mountain would follow the earth's curvature and glide in a circular path around the earth again and again (provided the cannoneer and the cannon got out of the way). Fired slower, the cannonball would strike the earth's surface; fired faster it would overshoot a circular orbit as we will discuss shortly. Newton calculated the speed for circular orbit, and since such a cannon-muzzle velocity was clearly impossible, he did not foresee humans launching satellites (and also because he probably didn't consider multi-stage rockets).

Note that in circular orbit, the speed of a satellite is not changed by gravity: only the direction changes. We can understand this by comparing a satellite in circular orbit with a bowling ball rolling along a bowling alley. Why doesn't the gravity that acts on the bowling ball change its speed? Because gravity is not pulling forward or backward; gravity pulls straight downward. The bowling ball has no component of gravitational force along the direction of the alley.

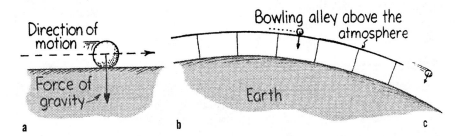

Consider a bowling alley that completely surrounds the earth, elevated high enough to be above the atmosphere and air drag. Not high enough to be "beyond earth gravity," as is commonly thought, for earth's gravity extends throughout the universe in accord with the inverse-square law. Many people

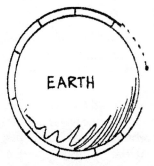

mistakenly think that satellites are above gravity. To repeat for emphasis, this simply is not so. What satellites are above is the atmosphere and air drag—*not* gravity!

A bowling ball will roll at constant speed along the alley. If a part of the alley is cut away, the ball would roll off its edge and hit the ground below. If the ball rolls faster along the alley, when it encounters the gap it will hit the ground farther along the gap. Is there a speed whereby the ball will clear the gap (like a motorcyclist who drives off a ramp and clears a gap to meet a ramp on the other side)? The answer is yes: 8 km/s will clear that gap—and any gap—even a 360° gap. It would be in circular orbit.

Note that a satellite in circular orbit is always moving in a direction perpendicular to the force of gravity that acts on it. The satellite does not move in the direction of the force, which would increase its speed, nor does it move in a direction against the force, which would decrease its speed. Instead, the satellite moves at right angles to the gravitational force that acts on it. With no component of motion along this force, no change in speed occurs—only change in direction. So we see why a satellite in circular orbit glides parallel to the surface of the earth at constant speed.

Questions

Consider a ball rolling along a bowling alley elevated above the atmosphere that completely circles the earth.

1. Why wouldn't the force of gravity change the speed of the ball?

2. If a section of the alley were cut away to leave a large gap, how fast must the ball travel to clear the gap and continue its motion as usual? At this speed, what would be the maximum gap for unchanged motion?

Putting a payload into earth orbit requires control over the speed and direction of the rocket that carries it above the atmosphere. A rocket initially fired vertically is intentionally tipped from the vertical course. Then, once above the drag of the atmosphere, it is aimed horizontally, whereupon the

---

Answers

1. A change in speed requires a force or component of force along the direction of travel. In this case the alley and gravitational force on the ball are everywhere perpendicular to each other, so there is no force component in the direction of motion.

2. To completely clear the gap, the ball must have orbital speed. Then its curved path matches that of the alley's surface. In this case the alley can be completely removed, which in effect is a 360° gap, because the ball would be in earth orbit anyway!

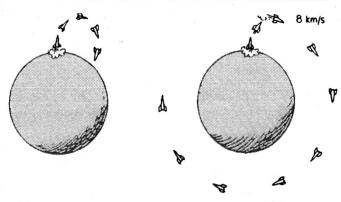

payload is given a final thrust to orbital speed. We see this in the sketch, where for the sake of simplicity the payload is the entire single-stage rocket. With the proper tangential velocity it falls around the earth, rather than into it, and becomes an earth satellite.

For a satellite close to the earth, the period (the time for a complete orbit about the earth) is about 90 minutes. For higher altitudes, the orbital speed is less and the period is longer. For example, communication satellites located in orbit 5.5 earth radii above the surface of the earth have a period of 24 hours. This period matches the period of daily earth rotation. For an orbit around the equator, these satellites stay above the same point on the ground. The moon is even farther away and has a period of 27.3 days. The higher the orbit of a satellite, the less its speed and the longer its period.[*]

Elliptical Orbits

If a projectile just above the drag of the atmosphere is given a horizontal speed somewhat greater than 8 kilometers per second, it will overshoot a circular path and trace an elliptical path.

An **ellipse** is a specific curve: the closed path taken by a point that moves in such a way that the sum of its distances from two fixed points (called *foci*) is constant. An ellipse can be easily constructed by using a pair of tacks, one at each focus, a loop of string, and a pencil. The closer the foci are to each other, the

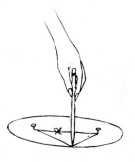

closer the ellipse is to a circle. When both foci are together, the ellipse is a circle. So we see that a circle is a special case of an ellipse. For a satellite orbiting earth, one focus is at the earth's center; the other focus could be inside or outside the earth.

Unlike the constant speed of a satellite in a circular orbit, speed varies in an elliptical orbit. When the initial speed is greater than 8 kilometers per sec-

---

[*] The speed of a satellite in circular orbit is given by $v = \sqrt{GM/d}$ and the period of satellite motion is given by $T = 2\pi\sqrt{d^3/GM}$, where G is the universal gravitational constant, M is the mass of the earth (or whatever body the satellite orbits), and d is the altitude of the satellite measured from the center of the earth or parent body.

Questions

1. If you've ever watched the launching of a satellite, you may have noticed that the rocket starts vertically upward, then departs from a vertical course and continues its climb at an angle. Why?

2. One of the beauties of physics is that there are usually different ways to view and explain a given phenomenon. Is the following explanation valid? Satellites remain in orbit instead of falling to the earth because they are beyond the main pull of earth's gravity.

3. Satellites in close circular orbit fall about 5 m during each second of orbit. Why doesn't this distance accumulate and send satellites crashing into the earth's surface?

ond, the satellite overshoots a circular path and moves away from the earth, against the force of gravity. It therefore loses speed. Like a rock thrown into the air, it slows to a point where it no longer recedes and then begins to fall

Answers

1. The initial vertical climb lets the rocket get through the denser atmosphere most quickly and is also the best direction at low initial speed, when a large part of the rocket's thrust is needed just to support the rocket's weight. Then the rocket must acquire enough tangential speed for orbit, which is why it departs from a vertical course until finally its path is horizontal for a circular orbit.

2. No, no, a thousand times no! If any moving object were beyond the pull of gravity, it would move in a straight line and would not curve around the earth. Satellites remain in orbit because they are being pulled by gravity, not because they are beyond it. For the altitudes of most earth satellites, the force of gravity is only a few percent weaker than at the earth's surface.

3. The falling distance below a straight-line tangent *does* accumulate with time. Satellite motion is easier understood by considering a *new* tangent line each second, where the 5-m fall procedure repeats itself each second. The process of falling with the curvature of the earth continues from tangent line to tangent line, so the curved path of the satellite and the curve of the earth's surface "match" all the way around the earth. Satellites in fact do crash to the earth's surface from time to time, but this is principally because they encounter air resistance in the upper atmosphere that decreases their orbital speed.

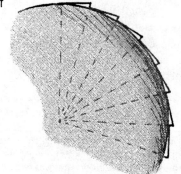

back toward the earth. The speed it loses in receding is regained as it falls back toward the earth, and it finally rejoins its original path with the same speed it had initially. The procedure repeats over and over, and an ellipse is traced each cycle.

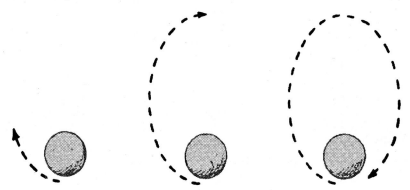

It's interesting to note that the curved path of a projectile such as a tossed baseball or a cannonball is actually a tiny segment of an ellipse that extends within and just beyond the center of the earth. Note in the sketch at right the several paths of cannonballs fired from Newton's mountain. All ellipses have the center of the earth as one focus. As

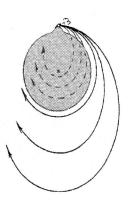

muzzle velocity is increased, the ellipses are less eccentric (wider), and when muzzle velocity reaches 8 kilometers per second, the ellipse rounds into a circle and does not intercept the earth's surface. The cannonball is then in circular orbit. At greater muzzle velocities, the orbiting cannonball traces the familiar external ellipse. It's interesting to note that for internal ellipses, the center of the earth is the far focus of the ellipse. For external ellipses, the earth's center plays the role of the near focus. The other focus has no special location and varies with the eccentricity of the ellipse.

Comets follow highly elliptical (very eccentric, stretched out) paths around the sun. Their orbits extend far beyond the planets and their orbital planes may be very different from the orbital planes of the planets. A typical comet is a chunk of ice and other materials that measure a few kilometers across. As a comet approaches the sun, solar heat vaporizes part of the ices that compose it,

Question

The orbital path of a satellite is shown in the sketch. In which of the marked positions A through D does the satellite have the greatest speed? Lowest speed?

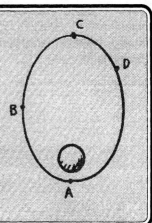

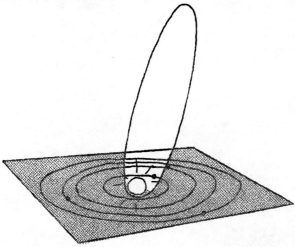

producing a fuzzy luminous ball of glowing vapors. Solar wind and radiation pressure blow vapors outward, away from the sun, into long flowing tails. A comet's tail can extend over 100 million kilometers. The densities of materials in a tail, interestingly, are less than the densities achieved here on earth in typical industrial vacuums. So compared with the atmosphere, the tail of a comet is "nothing at all." When the tail of a comet crosses the earth directly, except for meteor showers high in the atmosphere, nothing changes at the earth's surface. The impact of a comet nucleus, however, is a different story, as occurred so dramatically when Comet Shoemaker-Levy crashed into Jupiter in 1994. Craters are formed by the impact of comets as well as meteorites. Only the impact debris indicates the difference.

Answer

The satellite has its greatest speed as it whips around A and has its lowest speed at position C. Beyond C it gains speed as it falls back to A to repeat its cycle.

Comets are plentiful. There is almost always a comet in the sky, but most are too faint to be seen without a good telescope. Hale-Bopp in 1997 was a wonderful exception. About half a dozen new comets are discovered each year, mostly by amateur astronomers. Most comets have no visible tails, for their supply of ice is eventually exhausted. After about 100 to 1000 passes around the sun, a comet is pretty well sublimated away.

I can't think of comets without thinking of my dear friend and former student, Tenny Lim. In the early 80s I was successful in convincing her to pursue a career in science and engineering. Tenny, very good with her hands and her mind, had just graduated in dental technology. Nevertheless she began her studies all over again, spending four years rather than two years at City College of San Francisco, graduating and then going on to Cal Poly at San Luis Obispo. Today Tenny is part of an exciting team of engineers and scientists at Jet Propulsion Labs in Pasadena who are on a mission to find out what makes up the nucleus of a comet. The project, Deep Space 4/Champollion, is scheduled for launch in April-May 2003, to intercept periodic Comet Tempel 1. This will be the first landing of scientific instruments on the surface of a cometary nucleus. Funding permitting, extra-terrestrial samples may be collected and returned to a carrier spacecraft, and then to Earth. The samples would return to Earth in 2010 for analysis in terrestrial laboratories. I'm so proud of Tenny Lim!

Kepler's Laws of Planetary Motion

Newton's law of gravitation was preceded by three important discoveries about planetary motion by the German astronomer, Johannes Kepler, junior assistant to the famed Danish astronomer, Tycho Brahe. Brahe headed the world's first great observatory in Denmark, just before the advent of the telescope. Using huge brass protractor-like instruments called quadrants, Brahe measured the positions of planets over twenty years so accurately that his measurements are still used today. From Brahe's data, entrusted to Kepler after Brahe's death, Kepler discovered that the planets follow elliptical paths about the sun. This is Kepler's first law of planetary motion: The orbits of the planets describe ellipses, with the sun located at one focus of each ellipse. Kepler further discovered from Brahe's data that planets do not go around the sun at a uniform speed but move faster when they are nearer the sun and more slowly when they are farther from the sun. They do this in such a way that an imaginary line or spoke joining the sun and the planet sweeps out equal areas of space in equal times. We see in the figure that the triangular-shaped area swept out during a month when a planet is orbiting far from the sun (triangle ASB) is equal to the triangular area swept out during a month when the planet is orbiting closer to the sun (triangle CSD). This sweeping out of equal areas in equal times is Kepler's second law.

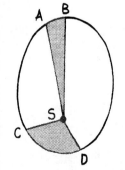

Kepler had no clear idea as to *why* the planets moved as they do. He was the first to coin the word satellite, but didn't see that a satellite is simply a projectile under the influence of a gravitational force directed toward the body the satellite orbits. He lacked a conceptual model. You know that if you toss a rock upward, it goes slower the higher it rises because it's going against gravity. And you know that when it returns it's going with gravity and its speed increases. Kepler didn't see that a satellite behaves the same way. Going away from the sun, it slows. Going toward the sun, it speeds up. A satellite, whether a planet orbiting the sun, or today's satellites orbiting the earth, moves slower going against the gravitational field and faster going with the field. Kepler didn't see this simplicity, and instead fabricated complex systems of geometrical figures to find sense in his discoveries. These proved futile. Ten years after the discovery of his two laws, with an enormous effort of numerical trial and error, Kepler came up with a third law of planetary motion. Kepler's third law is a mathematical relationship between a planet's distance from the sun and its period of orbit (distance cubed ~ period squared). Kepler's third law was inspiration in its time because it revealed the mathematical fabric of the solar system. It also proved to be a powerful tool, for if you know the radius of a planet's orbit, you can derive the radii of all the other planets from their orbital periods. But Kepler had no understanding of his mathematical relationship. It was obvious later to Newton. Today in physics classes, a routine exercise is having students equate Newton's law of gravitation to the equation for centripetal force $(G\frac{mM}{d^2} = \frac{mv^2}{d})$. Guess what emerges: that's right— Kepler's third law!

Energy Conservation and Satellite Motion

Recall from Chapter Five that a moving object has kinetic energy (KE) by virtue of its motion. An object above the earth's surface has potential energy (PE) by virtue of its position. Everywhere in its orbit, a satellite has both KE and PE with respect to the body it orbits. The sum of the KE and PE will be a constant all through the orbit. The simplest case occurs for a satellite in circular orbit.

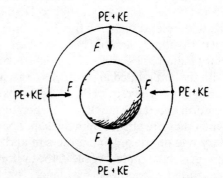

In circular orbit the distance between the body's center and the satellite does not change, which means the PE of the satellite is the same everywhere in orbit. Then, by the conservation of energy, the KE must also be constant. So a satellite in circular orbit coasts at an unchanging PE, KE, and speed.

In elliptical orbit, the situation is different. Both speed and distance vary. PE is greatest when the satellite is farthest away (at the apogee) and least when the satellite is closest (at the perigee). Note that the KE will be least when the PE is most, and the KE will be most when the PE is least. At every point in the orbit, the *sum* of KE and PE is the same.

This component of force does work on the satellite

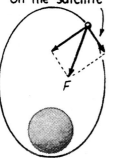

At all points along the elliptical orbit (except at the apogee and perigee) there is a component of gravitational force parallel to the direction of motion of the satellite. This component of force in the direction of motion changes the speed of the satellite. By the work-energy relationship, we can say this component of force × distance moved = ΔKE. Either way, when the satellite gains altitude and moves against this component, its speed, and KE, decrease. The decrease continues to the apogee. Once past the apogee, the satellite moves in the same direction as the component, and the speed and KE increase. This increase continues until the satellite whips past the perigee and repeats the cycle.

> **Question**
>
> In summary, why does the force of gravity change the speed of a satellite when it is in an elliptical orbit, but not when it is in a circular orbit?

Escape Speed

We know that a cannonball fired horizontally at eight kilometers per second from Newton's mountain would find itself in orbit. But what would happen if the cannonball were instead fired at the same speed *vertically*? It would

> **Answer**
>
> If there exists a component of force tangent to the path of the satellite, then acceleration will involve a change in speed as well as direction. In circular orbit the gravitational force is always perpendicular to the satellite's direction of motion, just as the radius of a circle is everywhere perpendicular to the circumference. So there is no component of gravitational force along the tangent, and only the direction of motion changes—not the speed. But when the satellite moves in directions that are not perpendicular to the force of gravity, as in an elliptical path, there is a component of force tangent to the path and the speed changes. A component of force tangent to the satellite's path does work to change its KE.

rise to some maximum height, reverse direction, and then fall back to earth. Then the old saying "What goes up must come down" would hold true, just as surely as a stone tossed skyward will be returned by gravity (unless, as we shall see, its speed is too great).

In today's space-faring age, it is more accurate to say "What goes up may come down," for there is a critical speed at which a projectile is able to outrun gravity and escape the earth. This critical speed is called **escape speed** or, if direction is involved, *escape velocity*. From the surface of the earth, escape speed is 11.2 kilometers per second.* Launch a projectile at any greater speed and it will leave the earth, traveling slower and slower due to earth gravity, never stopping. Gravitational interaction with the earth becomes weaker and weaker with increased distance, its speed becomes less and less, though both are never reduced to zero. The payload outruns the earth's influence. Although it never escapes the earth's gravitational field, it escapes the earth itself.

So the escape speed from the surface of Planet Earth is 11.2 km/s. The escape speeds from other bodies in the solar system are shown in the table below. Note that the escape speed from the sun is 620 kilometers per second

Escape Speeds at the Surface of Bodies in the Solar System			
Astronomical Body	Mass (number of earth masses)	Radius (number of earth radii)	Escape speed (km/s)
Sun	330,000	109	620
Sun (at a distance of the Earth's orbit)		23,500	42.2
Jupiter	318.0	11.00	60.2
Saturn	95.2	9.2	36.0
Neptune	17.3	3.47	24.9
Uranus	14.5	3.7	22.4
Earth	1.00	1.00	11.2
Venus	0.82	0.95	10.4
Mars	0.11	0.53	5.0
Mercury	0.05	0.38	4.3
Moon	0.01	0.27	2.4

---

* From an energy point of view, as a projectile continues outward, its PE increases and its KE decreases. By energy conservation, the PE of a 1-kg mass infinitely far from earth is 60 million joules. So to put a payload that far from the earth's surface requires an initial KE of at least 60 MJ/kg. This corresponds to a speed of 11.2 km/s, whatever the mass involved.

Escape speed, from any planet or any body, is given by $v = \sqrt{2GM/d}$, where G is the universal gravitational constant, M is the mass of the attracting body, and d is the distance from its center. (At the surface of the body, d would simply be the radius of the body.)

at the surface of the sun. Even at a distance from the sun equaling that of the earth's orbit, the escape speed from the sun is 42.2 kilometers per second, considerably more than the escape speed from the earth. An object projected from the earth at a speed greater than 11.2 kilometers per second but less than 42.2 kilometers per second will escape the earth, but not the sun. Rather than recede forever, it will take up an orbit around the sun.

Interestingly, escape speed might well be called the *maximum falling speed*. Any object, however far from earth, released from rest and allowed to fall to earth only under the influence of the earth's gravity, would not exceed 11.2 km/s. As a closer-to-home example, if you toss a ball upward at 30 m/s to a friend atop a building, and the ball just barely reaches your friend, then when the friend drops the ball back to you it will return at 30 m/s. So if it takes 11.2 km/s to launch a package to a friend beyond Pluto, you'll catch it at the same speed if it is dropped back to you.

The first probe to escape the solar system, Pioneer 10, was launched from earth in 1972 with a speed of only 15 kilometers per second. The escape was accomplished by directing the probe into the path of oncoming Jupiter. It was whipped about by Jupiter's great gravitation, picking up speed in the process—similar to the increase in the speed of a ball encountering an oncoming bat when it departs from the bat. Its speed of departure from Jupiter was increased enough to exceed the sun's escape speed at the distance of Jupiter. Pioneer 10 passed the orbit of Pluto in 1984. Unless it collides with another body, it will wander indefinitely through interstellar space. Like a note in a bottle cast into the sea, Pioneer 10 contains information about the earth that might be of interest to extraterrestrials, in hopes that it will one day wash up and be found on some distant "seashore."

It is important to point out that the escape speeds for different bodies refer to the initial speed given by a *brief* thrust, after which there is no force to assist motion. One could escape the earth at any *sustained* speed more than zero, given enough time. For example, suppose a rocket is launched to a destination such as the moon. If the rocket engines burn out when still close to the earth, the rocket needs a minimum speed of 11.2 kilometers per second. But if the rocket engines can be sustained for long periods of time, the rocket could go to the moon at any speed without ever attaining 11.2 kilometers per second.

Interestingly, the accuracy with which an unmanned rocket reaches its destination is not accomplished by staying on a pre-planned path or by getting back on that path if it strays off course. No attempt is made to return the rocket to its original path. Instead, the control center in effect asks, "Where is it now relative to where it ought to go? What is the best way to get there from here, given its present situation?" With the aid of high-speed computers, the answers to these questions are used in finding a new path. Corrective thrusters put the rocket on this new path. This process is repeated over and over again all the way to the goal.

Questions

1. If a flight mechanic drops a wrench from a high-flying jumbo jet, it crashes to earth. But if an astronaut outside the orbiting space shuttle drops a wrench, why doesn't it crash to earth also?

2. The orbiting space shuttle travels at 8 km/s around the earth. Suppose it projects a capsule rearward at 8 km/s relative to the shuttle. What path will the capsule take?

3. Orbital speed of the earth about the sun is 30 km/s. If the earth suddenly stopped in its tracks, what would occur?

4. If you wanted to send radioactive waste by rocket to the sun, how fast and in what direction relative to the earth's orbit should the rocket be fired?

5. Why don't communication satellites that "hover motionless" above the same spot on earth crash into the earth?

I ask my class if there's a lesson to be learned here. Suppose, for example that you find that you are "off course." You may, like the rocket, find it more fruitful to take a course that leads to your goal as best plotted from your present position and circumstances, rather than try to get back on the course you plotted from a previous position and under, perhaps, different circumstances. So many ideas in physics, it seems, have a moral.

Answers

1. If a wrench or anything else is "dropped" from an orbiting space vehicle, it has the same tangential speed as the vehicle and remains in orbit. If a wrench is dropped from a high-flying jumbo jet, it too has the tangential speed of the jet. But this speed is insufficient for the wrench to orbit the earth. Instead it soon crashes to the earth's surface.

2. When a capsule is projected rearward at 8 km/s relative to the shuttle, which is itself moving forward at 8 km/s relative to the earth, the speed of the capsule relative to the earth will be zero. It will have no tangential speed for orbit, so its path will simply be a vertical straight line to earth, where it will crash.

3. Like the capsule of the previous question, the earth would fall straight into the sun.

4. Fire the rocket at 30 km/s in a direction opposite to the earth's path around the sun. Then it will simply drop into the sun.

5. Although the velocity of a communications satellite is zero relative to a stationary spot on earth, its tangential velocity relative to the earth's center of mass is non zero, and is enough to keep it orbiting the earth rather than crashing to earth. Communication satellites appear motionless because their orbital period coincides with the daily rotation of the earth.

To date, rockets have been the carriers of launched satellites. Sometime in the future we may power rockets with nuclear fusion, but for the present, chemical rockets will have to do. The big problem with rockets is that they have to carry their own fuel and carry additional fuel to lift that fuel. So most of a rocket is fuel. Some 96% of the classic Saturn V rocket was its fuel and fuel tank. Before Saturn V had traveled even its own length, some 36 stories high, it burned a greater mass of fuel than the mass of the payload it delivered to the moon. Other vehicles, such as electric locomotives, don't carry fuel at all. Their power is obtained via overhead lines or through rails. Magnetically levitated vehicles don't carry fuel either. Tomorrow's satellites similarly may be launched without carrying fuel. Acceleration can be accomplished in horizontal tunnels along electromagnetic tracks without combating gravity. When launch speed is reached, the tube may curve upward, perhaps within a mountain, so the payload exits at a speed either sufficient for orbit or a head-start speed for a secondary means of propulsion. Little or no fuel will be accelerated, and air drag can be reduced by evacuating much of the air before launch. It's safe to speculate that today's chemical rockets that fight gravity every step of the way will seem quaint to our grandchildren.

As our space-faring efforts carry us farther into space, we may more and more come to see earth as our local address—and the entire universe as our home.

9

On Science

The physics of the preceding chapters presented a "this is the way we see it" point of view. They didn't dwell on "this is how we came to see it this way." More emphasis was placed on content than process. Many science types argue that the process of science is a more important lesson than science content. I feel both are essential, although I see the current emphasis in education on process sometimes taking content for granted. I'm disturbed that educated people commonly believe that in collisions, the forces are different on each colliding body; that the earth pulls harder on the moon than the moon pulls on the earth; and that satellites are beyond the pull of gravity. And as mentioned earlier, most people attribute the earth's inner heat to pressure rather than radioactivity. That's why I emphasize content in my books and in my videos. I'll hope that the efforts of others who emphasize process will provide the reader with a good balance.

So let's conclude this first book on physics with what is customarily a beginning chapter—a brief overview of science and scientific process. Science is the present-day equivalent of what used to be called *natural philosophy*, which was the study of unanswered questions about nature. As the answers were found, they became part of what is now called *science*. The study of science today branches into the study of living things and non-living things: the life sciences and the physical sciences. The life sciences branch into such areas as biology, zoology, and botany. The physical sciences branch into such areas as physics, chemistry, geology, meteorology, and astronomy. We're concerned with physics because it is basic to the other physical sciences. As we have seen, physics is about motion, force, energy, and the gravity that permeates the universe. In subsequent books we will see that it is also about matter, heat, sound, light, and the insides of atoms. Physics is the basic science, leading to chemistry, which tells us how matter is put together, how atoms combine to form molecules, and how those molecules combine to make the materials around us. Physics and chemistry applied to the earth and its processes make up the science of geology and, when applied to other planets and their stars, it is astronomy.

Biology is more complex than physics and chemistry, for it involves matter that is alive. Beneath biology is chemistry, and beneath chemistry is physics. The concepts of physics reach up to these more complicated sciences. Physics is basic to both physical science and life science. So we see that an understanding of science in general begins with an understanding of physics.

Nevertheless, in most high schools biology is taught first and, if at all, physics is taught last. This order has been set in stone for a hundred years, when biology was a classification science, mainly classifying plants and animals. Chemistry, before the electron had been discovered, was concerned with mixing chemicals. Physics was taught last because it was treated as applied mathematics. Even today, if you look in the table of contents of a biology book you see a lot of chemistry, and in the table of contents of a chemistry book you see a lot of physics. But in the table of contents of a physics book, there's no chemistry and no biology. It's ironic that the least complicated *course* in school may be biology, with the most complicated *course*, physics! One reason for this is the greater amount of mathematics usually embedded in a physics course.

Mathematics—The Language of Science

Mathematics is the core language of science—and particularly physics. When physics ideas are expressed mathematically, they are unambiguous and don't have the double meanings that so often confuse the discussion of ideas expressed in common language. Expressed mathematically, ideas are easier to verify or disprove by experiment. Research physicists communicate with one another primarily in mathematical language. An understanding of mathematics is essential to a mastery of physics.[*]

The intent of this book has not been to make you a master of physics,

---

[*] What do you think about more emphasis on probabilities in math classes? How about youngsters learning about simple probabilities and risk assessment, having some feeling for percentages and parts per million and parts per billion? Which is more useful in life, being able to sensibly assess relative dangers and benefits, or being able to do algebra? Shouldn't every educated citizen have a sound idea of what their odds are of winning a lottery?

but to familiarize you with some of its basic mechanics concepts. Although the connections of physics are presented in equation form, a mastery of equation manipulation isn't needed. We've seen that we can appreciate equations as short-hand abbreviations of the connections of nature without necessarily being able to carry out mathematical analysis. Equations show what is connected to what, and how they connect. Equations are short-hand statements and guides to thinking.

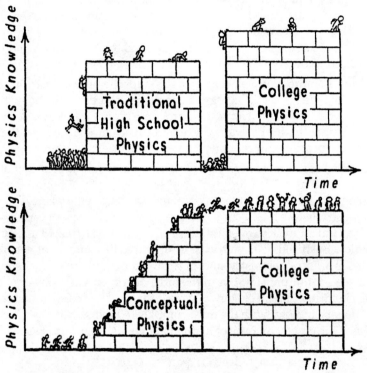

Measurements are a strong part of the laboratory portion of any physics course. Whichever course I teach, conceptual or otherwise, my first lecture concludes with a favorite mathematical exercise—measuring the diameter of the sun. I delight in holding up a meter stick and telling the class that with such a stick, a ruler, or a tape measure, they can determine the diameter of the sun. To prepare them for making this measurement, I first review how an image is made with a pinhole camera. I point out how the size of the image relates to the relative distances of object and image. I show that the ratio of distances of object and image from the pinhole is the same as the ratio of object size and image size.

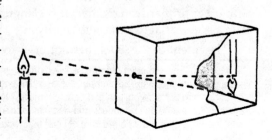

We're all familiar with the spots of sunlight in the shady area beneath trees on a sunny day. These spots of light are cast through the openings in leaves. But how many of us have noticed that these spots are circles—or ellipses, if the sun is low in the sky? That's because the spots of light are pinhole images of the sun! When the opening in the leaves is small compared with the distance to the ground below, the opening behaves as a pinhole in a pinhole camera. Just as ratios describe the relative sizes of image and object with a pinhole camera, similar ratios enable us to measure the sun's diameter. To facilitate measurement, I suggest that a piece of cardboard with a hole poked in its center be used instead of a tree. When held in the bright sunlight, the card will cast a shadow, and in the center of the shadow will be a circle—the image of the sun. If the sun were partially eclipsed, the image would be a crescent! The image of the sun is very small when the card is held close to the image. If the card is held farther from the shadow, the round solar image is larger. And the fascinating thing is that the ratio of the solar image diameter to the diameter of the sun is the same as the ratio of the distance between the image and the card and the distance to the sun. Given that they know that the sun is 93,000,000 miles or 150,000,000 kilometers distant, they're on their way to calculating its diameter. All they need to do is make the measurements cited in the practice page that I share with them.

The Language of Science in Words

Mathematical equations can be expressed in words as well as by symbols. In fact, if one can't express a mathematical equation in words, understanding is suspect. Many a student learns how to solve physics problems without understanding conceptually what is going on. Better physics courses now require that students provide explanations in words as well as with mathematics.

This book hasn't delved into mathematics. I have avoided mathematical manipulations, using equations only as guides to thinking and shorthand notation for nature's connections. But defining and using words requires as much care as defining and using mathematical symbols. Some central terms of physics are in the Physics Dictionary at the outset of this book: *fact, hypothesis, law, theory, concept,* and *prediction.* These terms often take on different meaning in science than outside of science. A fact, for example, may be taken to be an absolute in everyday language, but in science a fact does not pertain to an absolute—facts evolve. A hypothesis may mean speculation to the nonscientist, but to a scientist a hypothesis is an educated guess about nature, or a model of nature that seems to explain its laws. A hypothesis may be pretty much the same as a theory in everyday speech, but not in science. A theory in science is a broad set of

Facts are revisable data about the world. Theories interpret facts.

CONCEPTUAL **Physics** PRACTICE PAGE
Pinhole Image Formation

Look carefully at the round spots of light on the shady ground beneath trees. These are sunballs, and are actually images of the sun. They are cast by openings between leaves in the trees that act like the pinhole opening in a pinhole camera. Large sunballs, several centimeters in diam-

eter or so, are cast by openings that are relatively high above the ground, while small ones are produced by closer "pinholes." The interesting point is that the ratio of the diameter of the sunball to

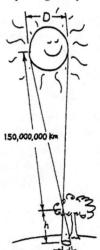

150,000,000 km

its distance from the pinhole is the same as the ratio of the sun's diameter to its distance from the pinhole. We know the sun is about 150,000,000 km from the pinhole, so careful measurement of this ratio tells us the diameter of the sun. That's what this page is about. Instead of finding sunballs under the shade of trees, make your own easier-to-measure sunballs.

1. Poke a small hole in a piece of cardboard (with a sharp pencil). Hold the cardboard in the sunlight and note the circular image that is cast. This is an image of the sun. Note that its size does not depend on the size of the hole in the cardboard, but only on its distance. The image will be a circle when cast on a surface that is perpendicular to the rays—otherwise it will be "stretched out" as an ellipse.

2. Try holes of different shapes; say a square hole, or a triangular hole. What is the shape of the image when its distance from the cardboard is large compared to the size of the hole? _____ Does the shape of the "pinhole" make a difference? _____

3. Measure the diameter of a small coin. Then place the coin on a viewing area that is perpendicular to the sun's rays. Position the cardboard so the image exactly covers the coin. Carefully measure the distance between the coin and the small hole in the cardboard. Complete the following:

$$\frac{\text{Diameter of sunball}}{\text{Distance to pinhole}} = \underline{\hspace{2cm}}$$

With this ratio, estimate the diameter of the sun.

WHAT SHAPE DO SUNBALLS HAVE DURING A PARTIAL ECLIPSE OF THE SUN?

Hewitt
Drawit!

ideas and equations, encompassing well tested hypotheses and laws.

A law of nature describes how nature behaves, as found from experience, and is *descriptive*, <u>not</u> *prescriptive*. The law of energy conservation, for example, summarizes a great many observations by saying that energy, although it can be transformed, is never created or destroyed. This is a statement about the way nature *does* behave, not the way it "should" behave. By contrast, a law setting a speed limit in a school zone is prescriptive. It tells us how we should behave. Laws of nature may be valid only within a limited area of experience. A law may turn out to be valid over a larger domain than was first imagined, yet not a limitless domain. Newton's law of gravitation, for example, is sufficient to get men to the moon, but is insufficient in describing nature where gravitation is extreme, as in a black hole.

A useful theory is able to predict how nature behaves in realms not yet explored. Prediction in science is not foretelling a future that is different from the past. It is foretelling the results of an experiment that will bring to light some process that may have occurred many times in the past but gone unobserved. A stockbroker predicts next week's market, which is sure to differ from all past markets. A scientist may predict what will happen when an electron with some specified great energy strikes a proton, something that has no doubt occurred often in the past but has perhaps never been observed. (Sometimes a scientist may predict what has never happened before, such as parachuting a spacecraft onto the surface of Mars, but, according to the theory that is applied, it *could* have happened before.)

Truth is a word seldom used in science (it seems to me that those who say they seek only "the truth" are more often seeking confirmation of what they already hold as true, and anything found that contradicts rather than supports, is discarded). Concepts are at the heart of what this book is about; the fabric that comprises theories. We've learned about the concepts of motion, acceleration, interactions, momentum and energy conservation, rotational motion, gravity, and satellite motion. As we have seen, concepts can be linked and explained with both mathematics and words.

The Scientific Method

One of the first things a student learns about science is The Scientific Method—the recipe by which scientists make many of their discoveries. This method goes back to founders in the sixteenth century, most notably, Galileo. The scientific method, extremely effective in gaining, organizing, and applying new knowledge, is essentially as follows:

1.   Recognize a problem.
2.   Make an educated guess—a **hypothesis.**
3.   Predict the consequences of the hypothesis.
4.   Perform experiments to test predictions.

5.   Formulate the simplest general rule that organizes the three main ingredients—hypothesis, prediction, and experimental outcome.

Although this method has a certain appeal, only occasionally has it been the key to the discoveries and advances in science. More often than not, trial and error, experimentation without guessing, and just plain accidental discovery account for much of the progress in science. The success of science has more to do with an *attitude* common to scientists than with a particular method. This attitude is one of inquiry, experimentation, and humility before the facts.

Ideally, scientists must accept their experimental findings even when they would like them to be different. They must strive to distinguish between what they see and what they *wish* to see, for scientists, like most people, have a vast capacity for fooling themselves. One of the first footnotes in my textbooks is "In your education it is not enough to be aware that other people may try to fool you; it is more important to be aware of your own tendency to fool yourself." Self-skepticism is better appreciated by those who have been completely convinced of something and who then find they are completely wrong. I recall the time I knew a particular friend was at a certain place with a certain other person. I knew this to be so—completely so. But later I found that both were in different countries at the time—that I was completely mistaken. After this experience of finding I was completely wrong, I've been less prone to fool myself. Less fortunate are those for whom similar experiences haven't happened or have escaped their notice, for they are more prone to question everything but their own knowledge. People have always tended to adopt general rules, beliefs, creeds, ideas, and hypotheses without thoroughly questioning their validity, and to retain them long after they have been shown to be meaningless, false, or at least questionable. And aren't the most widespread assumptions the ones least questioned? Too often, when an idea is adopted, particular attention is given to cases that seem to support it, while cases that seem to refute it are distorted, belittled, or ignored.

I was a child when World War II was being fought. I knew that on the battlefields of Europe, Germans on one side were fighting Americans on the other—to the death. I was puzzled that all the Germans held one point of view worth dying for, while all the Americans held another point of view, presumably opposite, worth dying for. I asked my mother how it could be that all the Germans and all the Americans could stand so differently on issues worth killing and dying for. It seemed to me there should be many Germans who would be inclined to believe what Americans stood for, and many Americans inclined to believe what the Germans stood for. It was then that I learned that right and wrong are relative terms. By definition, we're right and they're wrong. Few of us question what we've been taught to believe.

We also resist change. But the theories of science undergo change—they aren't fixed. Changes and advances in science don't take place by throwing out current ideas and techniques, but by pushing them to reveal new implications. Einstein, for example, never suggested that proven laws of physics be thrown out. Instead, he showed the laws of physics implied something that hadn't been revealed before. Scientific theories evolve as they go through stages of redefinition and refinement. During the last one-hundred years, for example, the theory of the atom has been repeatedly refined as new evidence on atomic behavior has been gathered. Similarly, chemists have refined their view of the way atoms bond together to form molecules; geologists have refined the plate tectonics theory; and astronomers, armed with new data from the Hubble telescope, are presently sharpening their view of the universe. The refinement of theories is a strength of science, not a weakness. Many people feel that it is a sign of weakness to change their minds. Competent scientists must be open to change. They change their minds, however, only when confronted with solid experimental evidence to the contrary or when a conceptually simpler hypothesis forces them to a new point of view. (I'm speaking ideally here, for often some competent scientists, like so many people, stubbornly cling to a notion that they identify with when a wealth of evidence is against it.) Better science occurs for those who are honest in the face of fact.

Away from their profession, scientists may be no more honest or ethical than other people. But in their profession they work in an arena that puts a high premium on honesty. The cardinal rule in science is that all hypotheses must be testable—they must be susceptible, at least in principle, to being proved *wrong*. In science, it is more important that there be a means of proving an idea wrong than there be a means of proving it right. At first this may seem strange, for when we wonder about most things, we concern ourselves with ways of finding out whether they are true. So instead of "How can I prove I'm right?", the scientist asks, "If I'm wrong, how would I know?" This emphasis on determining possible wrongness distinguishes science from nonscience. If you want to distinguish whether a hypothesis is scientific or not, look to see if there is a test for proving it wrong. If there is no test for its possible wrongness, then the hypothesis is not scientific.

Consider Darwin's hypothesis that life forms evolve from simpler to more complex forms. This could be proved wrong if paleontologists found that more complex forms of life appeared before their simpler counterparts. Einstein hypothesized that light is bent by gravity. This might be proved wrong if starlight that grazed the sun (which can be seen during a solar eclipse) were undeflected from its normal path. There are means of showing that Darwin's and Einstein's hypotheses are wrong, so they are scientific hypotheses. As it turns out, less complex life forms are found to precede their more complex counterparts and starlight is found to bend as it passes close to the sun. So both hypotheses are supported. If and when a hypothesis or scientific claim is confirmed, it is regarded as useful and a stepping stone to further knowledge.

Consider the hypothesis "life exists on other planets." Is this hypothesis scientific? The answer is no, for the statement is *speculation*. Although recent evidence suggests prior life in meteorites from Mars, there is no way to prove the hypothesis wrong if life weren't found.

Consider the hypothesis "intelligent civilizations exist elsewhere in the universe." This hypothesis is also not scientific, for, reasonable or not, it is speculation. Although it can be proved correct by the verification of a single instance of intelligent civilization existing elsewhere in the universe, there is no way to prove the hypothesis wrong if such is never found. If we search the far reaches of the universe for eons and do not find other civilizations, we could not prove that they don't exist around the next corner. A hypothesis that is capable of being proved right, but not capable of being proved wrong, is not a scientific hypothesis. Many such statements are quite reasonable and useful, but they lie outside the domain of science.

The requirement of a test for wrongness helps to sort out unscientific hypotheses that claim to be scientific. Consider the claim of a partygoer, that planets affect our character. Unless the partygoer can cite a test to show that planets don't affect our characters, hypothesizing they don't, the claim is unscientific. The claim may be true or it may be false—but in either case it is still unscientific. More important these days is the creationist theory, which has no test for wrongness. If the universe were not created by a supreme being, how could we know? Unless there is some test for this, the creationist hypothesis lies outside the domain of science—no matter how fascinating or important it may be.

The success of a scientific hypothesis rests not on its beauty, not upon the sincerity with which it is made, not upon the hopes or promises it implies—but on whether or not it stands up to *experiment*—reproducible by investigators in any culture and in any part of the world. Science, like mathematics, has no national language—it is transnational. Like a tree that grows ring by ring, scientific knowledge gained by experiment builds on prior knowledge. The built-in error-correcting methods of science produce a body of knowledge that helps us to have a better understanding of ourselves and the world and to shape a safer course for the future.

Experiment requires a patient attitude, for the hypotheses of most scientists fail the test of experiment. Benjamin Franklin said of experimentation, "In going on with these experiments, how many pretty systems do we build, which we soon find ourselves obliged to destroy?" He saw that the experience of experimenting suffices "to make a vain man humble." So whether your activities are scientific or not, when something doesn't work, don't spend your life pretending it does, or worse, trying to convince others of it.

No one has the time, energy, or resources to test every idea, so most of the time we take somebody's word. How do we know whose word to take? To reduce the likelihood of error, scientists accept the word only of those whose ideas, theories, and findings are testable—if not in practice, then at least in principle. Again, speculations that cannot be tested are regarded as unscientific. This has the long-run effect of compelling honesty—findings widely publicized among fellow scientists are generally subjected to further testing. Sooner or later, mistakes (or deception) are found out; wishful thinking is exposed. A discredited scientist doesn't get a second chance in the community of scientists. Honesty, so important to the progress of science, thus becomes a matter of self-interest to scientists. There is relatively little bluffing in a game where everyone can see the cards and all bets are called. In fields of study where falsehoods are not so easily revealed, the pressure to be honest is considerably less. A discredited governor today, for example, may be tomorrow's senator; a discredited physicist is out of the science arena altogether.

Question

Which of these is a scientific hypothesis?

a. Atoms are the smallest particles of matter that exist.

b. Space is permeated with an essence that is undetectable.

c. Albert Einstein is the greatest physicist of the twentieth century.

Pseudoscience

In pre-science times any attempt to harness nature meant forcing nature against her will. Nature had to be subjugated, usually with some form of magic or by means that were above nature—supernatural. Science does just the opposite and works within nature's laws. The methods of science have displaced the methods of magic—but not entirely. The old ways persist full force in primitive cultures, and they survive in technologically advanced cultures, too, often disguised as science. This is fake science—*pseudoscience.* The hallmark of a pseudoscience, or "alternate science," is that it lacks the key ingredients of evidence and having a test for wrongness. In the realm of pseudoscience, skepticism and tests for possible wrongness are downplayed or flatly ignored.

Answer

Only *a* is scientific because there is a test for falseness. The statement is not only *capable* of being proved wrong, but it in fact *has* been proved wrong. Statement *b* has no test for possible wrongness and is therefore unscientific. Likewise for any principle or concept for which there is no means, procedure, or test whereby it can be shown to be wrong (if it is wrong). Some pseudoscientists and other pretenders of knowledge will not even consider a test for the possible wrongness of their statements. Statement *c* is an assertion that has no test for possible wrongness. If Einstein was not the greatest physicist, how could we know? It is important to note that because the name Einstein is generally held in high esteem, it is a favorite of pseudoscientists. So we should not be surprised that the name of Einstein, like that of Jesus and other highly respected sources, is cited often by charlatans who wish to bring respect to themselves and their points of view. In all fields it is prudent to be skeptical of those who wish to credit themselves by calling upon the authority of others.

There are various ways to view our place in the universe, and mysticism is one of them. Astrology is an ancient form of magic that supposes a mystical connection between individuals and the universe—that human affairs are influenced by the positions and movements of stars and other celestial bodies. This non-scientific view is quite appealing with its implication that, however insignificant we may feel at times, we are in fact intimately connected to the workings of the cosmos—one created for humans, particularly those belonging to one's own tribe, community, or religious group. Astrology as ancient magic is one thing, but astrology in the guise of science is another. When it poses as a science related to astronomy, then it becomes pseudoscience. Some astrologers today present their craft in a scientific guise. When they use up-to-date astronomical information and computers that chart the movements of heavenly bodies, they're in the realm of science. But when they use this data to concoct astrological revelations, they have crossed over into full-fledged pseudoscience.

Pseudoscience, like science, makes predictions. The predictions of a dowser, who locates underground water with a dowsing stick, have a very high rate of success—nearly 100%. Whenever the dowser goes through his or her ritual and points to a spot on the ground, the well digger is sure to find water. Dowsing works. Of course, the dowser can hardly miss, because there is groundwater within 100 meters of the surface at nearly every spot on earth. (The real test of a dowser would be finding a place where water wouldn't be found!).

A shaman who studies the oscillations of a pendulum suspended over the abdomen of a pregnant woman can predict the sex of the fetus with an accuracy of 50%. This means if the shaman tries this magic many times on many fetuses, half the predictions will be right and half wrong—the predictability of ordinary guessing.

The best that can be said for the shaman is that the 50% success rate is a lot better than that of astrologers, palm readers, and other pseudoscientists who predict the future.

An example of pseudoscience that has zero success is provided by energy multiplying machines. These machines, which are alleged to deliver more energy than they take in, we are told, are "still on the drawing boards and needing funds for development." They are touted by quacks who sell shares to an ignorant public who succumb to the pie-in-the-sky promises of success. Then there are palm readers, who claim to have sacred knowledge of the "old ways" that transcends science. And others who claim possession of a "secret method" for astral travel, memory recovery, or channeling. Pseudoscientists are everywhere, are usually successful in recruiting apprentices for money or labor, and can be very convincing even to seemingly reasonable people. Their books greatly outnumber books on science in bookstores. Junk science is thriving.

We humans have learned much in the 500 years since the onset of science. Gaining this knowledge and overthrowing superstition required enormous human effort and painstaking experimentation. We should rejoice in what we've learned. We have come a long way in comprehending nature and in liberating ourselves from ignorance, and we should be proud of this. We no longer have to die whenever an infectious disease strikes. We no longer live in fear of demons. We no longer pour molten lead in the boots of women accused of witchery, as was done for nearly three centuries during medieval times. Today we have no need to pretend that superstition is anything but superstition, or that junk notions are anything but junk notions, whether dispensed by shamans, street-corner quacks, or hacks who write health books.

Yet there is reason to fear that what people of one time fight for, a following generation surrenders. The grip that belief in magic and superstition had on people took centuries to overcome. Yet today the same magic and superstition are perceived as enchanting to a growing number of people. James Randi reports in his book *Flim-Flam!* that more than twenty thousand practicing astrologers in the United States service millions of credulous believers. Science writer Martin Gardner reports that a greater percentage of Americans today believe in astrology and occult phenomena than did people in medieval Europe. Few newspapers carry a daily science column, but nearly all provide daily horoscopes. And then there are the flourishing television psychics who gain adherents daily.

Many people believe that the human condition is slipping backward because of growing technology. If we slip backward, however, it seems more likely it will be because science and technology will regrettably bow to the magic and superstitions of the past. Watch for their spokespeople. They're influential. Junk science is a huge and lucrative business.

Science, Art, and Religion

The search for order and meaning in the world has taken different directions: one is science, another is art, and another is religion. These three domains differ from one another in important ways, although they often overlap. Science is principally engaged with discovering and recording natural phenomena; the arts are an expression of human experience as it pertains to the senses; and religion addresses the source, purpose, and meaning of it all.

Science and the arts are comparable. In art we find what is possible in human experience. We can learn about emotions ranging from anguish to love, even if we haven't yet experienced them. The arts do not necessarily give us those experiences, but describe them to us and suggest possibilities. A knowledge of science similarly tells us what is possible in nature. Scientific knowledge helps us predict possibilities in nature even before these possibilities have been experienced. It provides us with a way of connecting things, of seeing relationships between and among them, and of making sense of the myriad of natural events around us. Science broadens our perspective of the natural environment of which we are a part. A knowledge of both the arts and the sciences makes for a wholeness that affects the way we view the world and the decisions we make about it and ourselves. A truly educated person is knowledgeable in both the arts and the sciences.

Science and religion have similarities also, but they are basically different from one another—principally because their domains are different. Science is concerned with the physical realm; religion is concerned with the spiritual realm. Simply put, science addresses *how*; religion addresses *why*. The practices of science and religion are also different. Whereas scientists experiment to find nature's secrets, religious practitioners worship their God and work to build human community. In these respects, science and religion are as different as apples and oranges and do not contradict each other. Science and religion are two different yet complementary fields of human activity.

Later, when we investigate the nature of light, we'll treat light first as a wave and then as a particle. To the person who knows only a little about science, waves and particles are contradictory; light can be only one or the other, and we have to choose between them. But to the enlightened person, waves and particles complement each other and provide a deeper understanding of light. In a similar way, it is mainly people who are either uninformed or misinformed about the deeper natures of both science and religion who feel that they must choose between a belief in religion or a belief in science. Unless one has a shallow understanding of either or both, there is no contradiction in being religious and being scientific in one's thinking.*

I vividly remember being approached by students while I was a guest lecturer in Bible Belt states who were painfully torn between "believing" in science which intrigued them, and believing in religion. They were enormously relieved when I told them that I thought people could embrace both science and religion without contradiction! When I cited the wave-particle example above, one student cried tears of joy, sobbing "Thank you" over and over. I went on to tell them that I was acquainted with scientists who didn't believe in a personal God, and scientists who were devoutly religious—and that both groups were generally happy citizens and first-rate scientists. And both groups included profoundly spiritual people. Einstein, who spelled God N-A-T-U-R-E and didn't believe in a personal God, put it well when he said "Science without religion is deaf; religion without science is blind." The religion and science he spoke of, of course, were deeper and more spiritual than the versions likely embraced by whomever was responsible for the pained confusion in the students mentioned above. How many young minds are stunted

---

* Of course, this doesn't apply to certain fundamentalists—Christian, Moslem, or otherwise— who steadfastly assert that one cannot embrace both their brand of religion and science.

by those who indoctrinate under the guise of education?

My beloved sister Marjorie Hewitt Suchocki is a highly spiritual person and a prominent Christian theologian at a graduate school in Claremont in southern California. Like many contemporary theologians, she views God as a supreme presence permeating all space and time. Her questions about meaning are in a realm related to, but distinct from, the realm of science; they do not conflict with science.

Science provides answers to questions that primarily ask *how*. Religion deals with *why*—questions about our place in the universe that have stirred the inquisitive souls of humans for centuries. A variety of religious answers abound, which gives richness to human diversity. An important message of science, however, is that not knowing is okay. It's okay not to know—especially in this awesome time of exploding knowledge about who and where we are, when an open mind is better prepared to discover *why* we are.

The tragic part of our past is replete with the power of those who claimed attainment of absolute certainty. The bright part of our future will be devoid of their influence. And if we find that one of nature's rules is that absolute certainty will always elude us, so be it. Again, *it's okay not to know*. This personal attitude widened to all fields of knowledge leads to wisdom.

Science and Technology

Science and technology are also different from each other. My friend Charlie Spiegel likes to say that a scientist knows why things should work; a technologist knows why they don't work—and a mathematician doesn't care! Whereas science involves discovering evidence and relationships for observable phenomena in nature and establishing theories that organize and make sense of it all, technology involves tools, techniques, and procedures for putting the findings of science to use.

We can never do just one thing. Doing this affects that, for there are side effects to our efforts. We are all familiar with the abuses of technology. Many people blame technology itself for widespread pollution, resource depletion, and even social decay in general—so much so that the promise of technology is obscured. That promise is a cleaner and healthier world. It is much wiser to combat the misuse of technology with knowledge than with ignorance. Wise applications of science and technology can lead to a better world.

Making a better world via art, science, and technology, was the personal goal of one of the most wonderful men I have ever met—Frank Oppenheimer. Frank founded the world's foremost museum linking art, science, and technology—the Exploratorium in San Francisco. Frank introduced hands-on exhibits that puzzled participants, and more importantly, that led to answers by personal interaction. Frank was interested in teaching people, especially young ones, to base answers to questions on personal observations. I began teaching at the Exploratorium in 1981, and was honored and delighted when Frank gave guest lectures, usually when we got to his favorite topics—the physics of light and of music. His method of teaching was to ask many more questions than he answered, prompting people to figure out their own answers. I loved and admired him, and dedicated the Fifth Edition of *Conceptual Physics* to him. Sadly, lung cancer claimed Frank a few days before the book was printed. He didn't know of the dedication, but wrote an opening paragraph for the book. Here is what he wrote:

"By trying to understand the natural world around us, we gain confidence in our ability to determine whom to trust and what to believe about other matters as well. Without this confidence, our decisions about social, political, and economic matters are inevitably based entirely on the most appealing lie that someone else dishes out to us. Our appreciation of the noticings and discoveries of both scientists and artists therefore serves, not only to delight us, but also to help us make more satisfactory and valid decisions and to find better solutions for our individual and societal problems."

Frank, like his brother Robert J. Oppenheimer, had good reason to cherish valid decision making. In the post-war years, both were victims of ill-informed decisions by government agents who hounded them for their involvement in the 1930s with communism. This was when the country was in the depths of the depression and concerned citizens were exploring alternatives to economic and political systems. During the hysteria of the cold war,

Answer

All of them! The human value of science, however, may be the least understood by most individuals in our society. The reasons are varied, ranging from the common notion that science is incomprehensible to people of average intellectual ability to the extreme view that science is a dehumanizing force in our society. Most of the misconceptions about science probably stem from the confusion between the *abuses* of scientific knowledge and scientific knowledge itself.

Building scientific knowledge is an enchanting human activity shared by a wide variety of people who, with present-day tools and know-how, are reaching further and finding out more about themselves and their environment than people in the past were ever able to do. The more you know about science, the more passionate you feel toward your surroundings. There is physics in everything you see, hear, smell, taste, and touch!

Frank was banished for ten years to a cattle ranch in Colorado, where he was forbidden to do physics research. Adding insult to injury, FBI agents admonished neighbors to beware of Frank and to watch for suspicious activities. Frank was deeply hurt by all this, and vowed to spend his remaining years teaching people to make informed decisions. Hence the Exploratorium.

Another renowned physicist of similar stature to Frank also gave guest lectures in my Exploratorium classes—Albert Baez. Albert wrote a physics textbook back in the 60s, *Physics—A Spiral Approach*, which, along with the books of Kenneth Ford, Eric Rogers, and Richard Feynman, influenced me to write my own. Al Baez's interests in physics, like Frank's, are mainly music and light. Al's passion for music precedes the success of his famous daughter, Joan. Al stresses what he calls the 4 C's: curiosity, creativity, competence, and compassion. Al himself is an example of each, especially the last. Problems facing us, he says, are given by the 4 P's: population, pollution, poverty, and proliferation of weapons of mass destruction.

In Perspective

Only a few centuries ago the most talented and most skilled artists, architects, and artisans of the world directed their genius and effort to the construction of the great cathedrals, synagogues, temples, and mosques. Some of these architectural structures took centuries to build, which means that nobody witnessed both the beginning and the end of construction. Even the architects and early builders who lived to a ripe old age never saw the finished results of their labors. Entire lifetimes were spent in the shadows of construction that to them were without beginning or end. This enormous focus of human energy was inspired by a vision that went beyond worldly concerns—a vision of the cosmos. To the people of that time, the structures they erected were their spaceships of faith, firmly anchored, but pointing to the cosmos.

Today the efforts of many of our most skilled scientists, engineers, artists, and artisans are directed to building the spaceships that already orbit the earth and others that will voyage beyond. The time required to build these spaceships is extremely brief compared to the time spent building the stone and marble structures of the past. Many people working on today's spaceships were alive before the first jetliner carried passengers. Where will younger lives lead in a comparable time?

We are at the dawn of a major change in human growth. This is not appreciated by those who benefit from the progressive uphill climb in health, nutrition, comfort, and human well-being over the past centuries. For too many, perception along the climb has been that advances have peaked—with only decline to come. Prophets of doom have never lacked public attention. Some welcome doom as fulfilling certain religious prophesies. But nevertheless, we advance. We are some 15 billion years from the Big Bang, with at least 100 billion more years to go. So like a child, we have more future than past.

Buckminister Fuller put it well when he likened us to chicken eggs about to be hatched. Inside the limited environment of the eggs are undeveloped chickens consuming and exhausting their inner-egg resources. It may seem to the chicks that doom is at hand, until in desperation the chicks break through their shells and hatch—entering a whole new range of possibilities. Similarly, the earth is our cradle and has served us well. But cradles, however comfortable, are one day outgrown. Like the chicks poking through the shell, we probe habitats beyond the earth. So with the inspiration that in many ways is similar to the inspiration of those who built the early cathedrals, synagogues, temples, and mosques, we aim for the cosmos.

We live in an exciting time!

index of names

general index